Foreword

You have in your hand the third *Official Guide* designed to help students prepare for examinations produced under the auspices of the Division of Chemical Education (DivCHED) of the American Chemical Society (ACS). The first guide, published in 1998 for general chemistry students, proved to be very popular as did the second guide, published in 2002 for organic chemistry students.

As is common for materials produced by the Exams Institute, we called on colleagues in the chemistry education community to help us put the guide together. A distinguished group of active, and respected, physical chemistry faculty members from all over the country accepted the challenge, rolled up their sleeves, and went to work. (A complete list of who did what appears on the *Acknowledgements* page.) This *Official Guide*, more than three years in the making, is the culmination of their wonderful *volunteer* efforts.

As a discipline, chemistry is surely unique in the extent to which its practitioners provide beneficial volunteer service to the teaching community. ACS exams have been produced by volunteer teacher-experts for more than seventy years. Other projects of the Examinations Institute benefit from the abundance of donated time, effort, and talent. The result is that invariably high quality chemistry assessment materials are made available to the teaching (and learning) community at a fraction of their real value.

The three *Official Guides* that have been released so far are intended to be ancillary student materials, particularly in courses that use ACS exams. As we noted in the Foreword to the general chemistry study guide, the care that goes into producing ACS exams may be lost on students who view the exams as foreign and unfamiliar. The purpose of this series of guides is to remove any barriers that might stand in the way of students demonstrating their knowledge of chemistry. The extent to which this goal is achieved in the physical chemistry study guide will become known only as future generations of chemistry students sit for an ACS exam in physical chemistry.

We wish them the best.

Thomas Holme
Kristen Murphy

Ames, Iowa
July, 2009

Acknowledgements

A work of this magnitude is the end result of significant efforts from a large number of people. The only way to describe our reaction to the dedication of the volunteers who made this study guide possible is that we are humbled. A topic as complex as physical chemistry is a challenge to summarize and communicate in a format like this one, and the committee of people who worked on this book did a superb job. The effort put forward was terrific, the results are compelling and the chance to work with such talented and enthusiastic colleagues is priceless. Everybody who pitched in has our heartfelt gratitude and to say "*Thanks, you did an outstanding job!*" only scratches the surface of our debt of gratitude.

Physical *Official Guide* Committee	
Richard Schwenz, Co-Chair	University of Northern Colorado
Theresa Julia Zielinski, Co-Chair	Monmouth University
Julie Boerio-Goates	Brigham Young University
Michelle Francl	Brynmawr College
Lynn Geiger	University of Colorado – Boulder
Dotie Sipowska	University of Michigan

This *Official Guide* also benefited from the careful proofreading by several colleagues. We extend our special thanks to these faculty members.

The personnel of the ACS Division of Chemical Education Examinations Institute played a central role in helping us to produce **Preparing for Your ACS Examination in Physical Chemistry: The Official Guide.** A very special thank you for all of the work involved is owed to our staff members.

While all of these reviewers have been very helpful in finding problems large and small, any remaining errors are solely our responsibility. You can assist us in the preparation of an even better product by notifying the Exams Institute of any errors you may find.

Thomas Holme
Kristen Murphy

Ames, Iowa
July, 2009

How to Use This Book

Students often express concerns about the nature of standardized, multiple-choice exams. In some cases they contend that questions are "tricky" or perhaps that they feel constrained by time and do poorly, relative to their knowledge of the subject. This book is designed to help students overcome these reactions to the ACS Physical Chemistry exams (we use "exams" because there are several forms for this particular sub-field of chemistry.)

The major divisions of this book correspond to the common groupings of topics covered by ACS exams for physical chemistry. The chapters themselves are collected into sections corresponding to the coverage of the three subject area exams, Thermodynamics, Dynamics (Kinetics) and Quantum Mechanics. Your instructor may be giving you an exam that is entirely from one of these sections. Alternatively, there are two Physical Chemistry exams that include questions from multiple sections. The Comprehensive exam is a single exam with questions from all three sections. There is also a "combined semester" exam that allows sections to be chosen to match the specific content covered in any given physical chemistry course. Be sure to check with your instructor about which exam is being used in your course.

Each topic group is introduced with a short discussion of the important ideas, concepts, and knowledge that are most frequently stressed in physical chemistry courses. *These discussions are not a substitute for studying your textbook, working the problems there, and discussing the challenging ideas with your teachers and fellow students.* Rather, they are reminders of what you have studied and how that fits into a larger understanding of that part of the natural world that we call chemistry.

Next, questions are presented that address those ideas. These questions have been drawn from past ACS exams or from items that were developed for exams but not used, and they should give you a good idea of the depth and range of understanding that is expected. Each question is dissected, and you will see how chemists think through each of the questions to reach the intended response. You will also see how choosing various wrong responses reveal misconceptions, careless computation, misapplication of principles, or misunderstandings of the material. Knowing how each incorrect answer is generated will assist you in diagnosing problems with your grasp of the principle being examined.

For physical chemistry, there are two types of questions, those that require calculators and those that do not. The set of exams that were released in 2006 were designed specifically to be completed without the need for a calculator. Older exams include items that use a calculator. Because this book is designed to be useful for any active physical chemistry exam, it includes both type of items. Again, you should check with your instructor to find out if you will be taking an exam that allows the use of a calculator or not.

This book is not designed to be a manual on test taking strategies. With a multiple choice format, there are multiple usable strategies, including to work the problem and find the correct choice, or to look at responses and eliminate incorrect responses. This book essentially works from the premise that the strategy with the highest probability for success is to know the correct answer and find that response.

The most effective way to use this book therefore, is to answer each **Study Question** before looking at the discussion of the item. Jot down a note of how you arrived at the answer you chose. Next, look at the analysis of the question. Compare your approach with that of the experts. If you missed the item, do you understand why? If you chose the correct response, was it based on understanding or chance? After you have spent time with the **Study Questions**, treat the **Practice Questions** as if they were an actual exam. Allow yourself 50 minutes, and write down your response to each question. Finally, score yourself. Go over the practice questions again. Write down what you needed to know before you could answer the question; and write down how you should think the problem through to reach the intended answer.

This book is designed to help you demonstrate your *real* knowledge of chemistry. When you take a physical chemistry examination prepared by the American Chemical Society Examinations Institute, you should be permitted to concentrate on demonstrating your knowledge of chemistry, and not on the structure of the examination. We sincerely hope that *The Official Guide* will enrich your study of chemistry, and minimize the trauma of effectively demonstrating what you have learned.

Sample Instructions

You will find that the front cover of an ACS Exam will have a set of instructions very similar to this. This initial set of instructions is meant for both the faculty member who administers the exam and the student taking the exam. You will be well advised to read the entire set of instructions while waiting for the exam to begin. This sample is from the physical chemistry exam released in 2006. Note that there are many forms of this exam, so the number of items and the time allowed can vary from the information here.

> **TO THE EXAMINER:** This test is designed to be taken with a special answer sheet on which the student records his or her responses. All answers are to be marked on this answer sheet, not on the test booklet. Each student should be provided with a test booklet, one answer sheet, and scratch paper; all of which must be turned in at the end of the examination period. The test is to be available to the students only during the examination period. For complete instructions, refer to the *Directions for Administering Examinations*. Calculators and Personal Digital Assistants are **not permitted**. Norms are based on:
>
> Score = number of right answers
> 50 items — 100 minutes
>
> **TO THE STUDENT:** DO NOT WRITE ANYTHING IN THIS BOOKLET! Do not turn the page until your instructor gives you the signal to begin.

Note the **restriction** on the **time** for administering the exam. This restriction applies to allow your results to be compared to national norms, ensuring that all students have had the same tools and time to display their knowledge. Your instructor may choose not to follow the time restriction, particularly if they do not plan to submit your data as part of the national process for calculating norms.

Be sure to notice that scoring is based **only** on the number of right answers. There is no penalty, therefore, for making a reasonable guess even if you are not completely sure of the correct answer. Often you will be able to narrow the choice to two possibilities, improving your odds at success. You will need to keep moving throughout the examination period, for it is to your advantage to attempt every question. Do not assume that the questions become harder as you progress through an ACS Exam. Questions are not grouped by difficulty, but by topic.

Next, here is a sample of the directions you will find at the beginning of an ACS exam.

> **DIRECTIONS**
> - When you have selected your answer, blacken the corresponding space on the answer sheet with a soft, black #2 pencil. Make a heavy, full mark, but no stray marks. If you decide to change an answer, erase the unwanted mark very carefully.
> - Make no marks in the test booklet. Do all calculations on scratch paper provided by your instructor.
> - There is only one correct answer to each question. Any questions for which more than one response has been blackened **will not be counted**.
> - Your score is based solely on the number of questions you answer correctly. **It is to your advantage to answer every question.**
> - The best strategy is to arrive at your own answer to a question before looking at the choices. Otherwise, you may be misled by plausible, but incorrect, responses.

Pay close attention to the mechanical aspects of these directions. Marking your answers without erasures helps to create a very clean answer sheet that can be read without error. As you look at your Scantron® sheet before the end of the exam period, be sure that you check that every question has been attempted, and that only one choice has been made per question. As was the case with the cover instructions, note that your attention is again directed to the fact that the score is based on the total number of questions that you answer correctly. You also can expect a reasonable distribution of **A**, **B**, **C**, and **D** responses, something that is not necessarily true for the distribution of questions in *The Official Guide*.

Sample of Given Information

Sign Convention
Work (w) is defined as having a positive (+) sign when work is done **on** the system by the surroundings.

Quantity	Symbol	Value	
internal energy	U		
enthalpy	H		
entropy	S		
Gibbs energy	G		
Helmholtz energy	A		
speed of light, vacuum	c	2.9979×10^8 m·s^{-1}	
Planck constant	h	6.6261×10^{-34} J·s	
elementary charge	e	1.6022×10^{-19} C	
electron mass	m_e	9.1094×10^{-31} kg	
proton mass	m_p	1.6726×10^{-27} kg	
atomic mass unit	u	1.6606×10^{-27} kg	
Rydberg constant	R_ν	1.09737×10^5 cm^{-1}	
Avogadro constant	N_A	6.02221×10^{23} mol^{-1}	
Faraday constant	F	96485.3 C·mol^{-1}	
ideal gas constant	R	8.3145 J·K^{-1}·mol^{-1}	1.987 cal·K^{-1}·mol^{-1}
		0.083145 L·bar·K^{-1}·mol^{-1}	0.082058 L·atm·K^{-1}·mol^{-1}
Boltzmann constant	k_B	1.3807×10^{-23} J·K^{-1}	

PERIODIC TABLE OF THE ELEMENTS

1	2	3	4	5	6	7	8	9	10	11	12	13	14	15	16	17	18
1 H 1.008																	2 He 4.003
3 Li 6.941	4 Be 9.012											5 B 10.81	6 C 12.01	7 N 14.01	8 O 16.00	9 F 19.00	10 Ne 20.18
11 Na 22.99	12 Mg 24.31											13 Al 26.98	14 Si 28.09	15 P 30.97	16 S 32.07	17 Cl 35.45	18 Ar 39.95
19 K 39.10	20 Ca 40.08	21 Sc 44.96	22 Ti 47.88	23 V 50.94	24 Cr 52.00	25 Mn 54.94	26 Fe 55.85	27 Co 58.93	28 Ni 58.69	29 Cu 63.55	30 Zn 65.39	31 Ga 69.72	32 Ge 72.61	33 As 74.92	34 Se 78.96	35 Br 79.90	36 Kr 83.80
37 Rb 85.47	38 Sr 87.62	39 Y 88.91	40 Zr 91.22	41 Nb 92.91	42 Mo 95.94	43 Tc (98)	44 Ru 101.1	45 Rh 102.9	46 Pd 106.4	47 Ag 107.9	48 Cd 112.4	49 In 114.8	50 Sn 118.7	51 Sb 121.8	52 Te 127.6	53 I 126.9	54 Xe 131.3
55 Cs 132.9	56 Ba 137.3	57 La 138.9	72 Hf 178.5	73 Ta 180.9	74 W 183.8	75 Re 186.2	76 Os 190.2	77 Ir 192.2	78 Pt 195.1	79 Au 197.0	80 Hg 200.6	81 Tl 204.4	82 Pb 207.2	83 Bi 209.0	84 Po (209)	85 At (210)	86 Rn (222)
87 Fr (223)	88 Ra (226)	89 Ac (227)	104 Rf (261)	105 Db (262)	106 Sg (263)	107 Bh (262)	108 Hs (265)	109 Mt (266)	110 Ds (269)	111 Rg (272)	112 Uub (277)		114 Uuq (2??)		116 Uuh (2??)		118 Uuo (2??)

58 Ce 140.1	59 Pr 140.9	60 Nd 144.2	61 Pm (145)	62 Sm 150.4	63 Eu 152.0	64 Gd 157.3	65 Tb 158.9	66 Dy 162.5	67 Ho 164.9	68 Er 167.3	69 Tm 168.9	70 Yb 173.0	71 Lu 175.0
90 Th 232.0	91 Pa 231.0	92 U 238.0	93 Np (237)	94 Pu (244)	95 Am (243)	96 Cm (247)	97 Bk (247)	98 Cf (251)	99 Es (252)	100 Fm (257)	101 Md (258)	102 No (259)	103 Lr (262)

Table of Contents

Foreword i

Acknowledgements ii

How to Use This Book iii

Sample Instructions iv

Sample of Given Information v

Table of Contents vi

Section 1 – Thermodynamics

    Equations of State 1

    Laws of Thermodynamics and State Functions 8

    Mathematical Relationships in Thermodynamics 15

    Chemical and Phase Equilibria 21

Section 2 – Dynamics

    Kinetic Molecular Theory 30

    Transport Properties 37

    Phenomenological Kinetics 43

    Mechanisms 50

    Reaction Dynamics 58

    Statistical Mechanics 65

Section 3 – Quantum Mechanics

    Quantum Chemistry: History and Concepts 71

    Simple Analytical Quantum Mechanical Model Systems 80

    Modern Quantum Mechanical Problems: Atomic Systems 90

    Symmetry 98

    Molecular Orbital Theory 104

    Spectral Properties 111

    Advanced Topics: Electronic Structure Theory and Spectroscopy 119

Equations of State

Chemistry is an experimental science, and the ability to describe chemical phenomena in terms of experimentally measurable variables is a key component of the field. For many systems, the relationship between physical variables provides a vital ability to infer information without measuring every possible physical variable. A mathematical relationship that characterizes the system (and state of the system) is referred to as an equation of state.

The equation of state relates the pressure, temperature and volume of a system. Observable effects of these variables are particularly relevant for gas phase systems because of the extent to which the state of a gas may change with changes in these variables.

The equation of state of an ideal gas is given by one of the more recognizable equations in all of science,
$$PV = nRT.$$
There are a number of simpler, qualitative relationships summarized by this equation, including the historical gas laws relating variables in pair-wise fashion. Thus, for example, Boyle's Law, which states that pressure and volume are inversely proportional for a gas is contained within this equation (by holding the temperature and amount of material constant.)

In many applications, however, the power of the equation of state arises not from the qualitative statements it makes but rather the calculations that can be performed using it. For gases, the ideal gas equation of state finds common usage in chemistry problems, but it is important to recognize that there are circumstances where gases do not behave ideally. In this chapter, equations of state for both, ideal and real gases, are examined with special attention to drawbacks of the ideal gas equation and the manner these drawbacks are corrected by empirical expressions.

Study Questions

ES-1. For an ideal gas, $Z = \dfrac{PV_m}{RT}$ is unity. Which is true in general for a real gas?

(A) $Z = 1$ for all pressures

(B) $Z < 1$ at low pressures and $Z < 1$ at high pressures

(C) $Z > 1$ at low pressures and $Z < 1$ at high pressures

(D) $Z < 1$ at low pressures and $Z > 1$ at high pressures

Knowledge Required:. Distance dependence of intermolecular forces

Thinking it Through: $Z = 1$ for all pressures only for an ideal gas and therefore response **(A)** is incorrect. Intermolecular forces existing between real gas molecules depend on intermolecular distances and forces. At low pressures the interparticle distances are large and attractive forces reduce the molar volumes relative to that of a perfect gas. Therefore at low pressures $Z < 1$ meaning that response **(C)** is incorrect. At high pressures the distances are short and the repulsive forces increase the molar volume and decrease the space available to each molecule causing $Z > 1$ eliminating response **(B)** and leaving response **(D)** as the correct choice.

Equations of State

ES-2. Gases are in corresponding states when they have the same reduced temperatures and pressures. Under what conditions is H_2 in the state corresponding to N_2 at 126 K and 1 atm?

Gas	T_c / K	P_c / atm
H_2	33	13
N_2	126	39

(A) 126 K, 1 atm
(B) 126 K, 39 atm
(C) 33 K, 3 atm
(D) 33 K, 0.33 atm

Knowledge Required:. Definition of reduced parameters in terms of critical parameters. Definition of the corresponding states.

Thinking it Through: According to the definition, the reduced temperature $T_r = \dfrac{T}{T_c}$ and the reduced pressure $P_r = \dfrac{P}{P_c}$. According to the information provided in this question, the reduced parameters for N_2 are $T_r = \dfrac{126\ \text{K}}{126\ \text{K}} = 1$ and $P_r = \dfrac{1\ \text{atm}}{39\ \text{atm}} = 0.026$. In order for H_2 to be in the state corresponding to N_2 at 126 K and 1 atm it must have the same reduced temperature and pressure, that is $T_r = 1$ and $P_r = 0.0256$. These values of reduced parameters allow you to determine that H_2 should be at $T = T_r \times T_c = 1 \times 33\ \text{K} = 33\ \text{K}$ and $P = P_r \times P_c = 0.026 \times 13\ \text{atm} = 0.33\ \text{atm}$. Response **(D)** is therefore correct. If you define corresponding states as states at the same temperature and pressure you will choose incorrect response **(A)**. Response **(C)** can be obtained under incorrect assumption that actual pressures of two gases at corresponding states should be at the same ratio as critical pressures.

ES-3. The valve between the 2.00-L bulb, in which the gas pressure is 1.00 atm, and the 3.00-L bulb, in which the gas pressure is 1.50 atm, is opened. What is the final pressure in the two bulbs, the temperature being constant and the same in both bulbs?

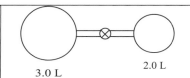

3.0 L   2.0 L

(A) 0.90 atm   (B) 1.25 atm   (C) 1.30 atm   (D) 2.50 atm

Knowledge Required: The ideal gas law.

Thinking it Through: According to the ideal gas equation of state, $PV = nRT$, it is important to notice that at constant temperature the number of moles is directly proportional to the product of pressure and volume, PV. After the valve is opened, the total volume is $(2.00 + 3.00)\ \text{L} = 5.00\ \text{L}$ and the total number of moles is the sum of moles initially in separate bulbs. Since both, volumes and number of moles add, we can say that

$$1.00\ \text{atm} \times 2.00\ \text{L} + 1.50\ \text{atm} \times 3.00\ \text{L} = P \times 5.00\ \text{L}$$

and therefore $P = 1.30$ atm, response **(C)** is correct. Assumption that the final pressure is an average of two separate pressures gives incorrect response **(B)**. Response **(D)** is just the sum of two pressures, not a correct approach. You will obtain response **(A)** if you incorrectly assume that at constant temperature PV is always constant.

ES-4. Which of these isotherms is experimentally observed near the critical temperature of a real gas?

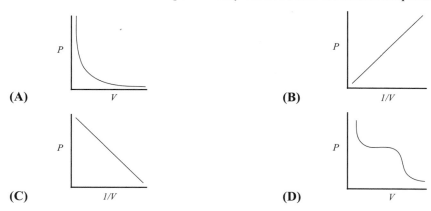

Knowledge Required: Graphical representation of gas law relationships, particularly for real gases.

Thinking it Through: Response **(A)** represents a hyperbolic graph. Our mathematics knowledge tells us this curve is PV = constant, a constant temperature and mass graph for an ideal gas which does not exhibit a critical point. Response **(B)** is the same as response **(A)** (PV = constant) but with a change in the horizontal axis. Response **(C)** gives P = constant – another constant / V which is not an applicable gas law. Response **(D)** shows the correct P-V behavior of a real gas near the critical point. The critical point itself occurs when $\left(\frac{\partial P}{\partial V}\right)_T = 0$ and $\left(\frac{\partial^2 P}{\partial V^2}\right)_T = 0$

ES-5. The van der Waals equation of state $\left(P + \frac{n^2 a}{V^2}\right)(V - nb) = nRT$ contains a term representing a "molecular size". The approximate magnitude of this term is

(A) 10^{-6} cm$^3 \cdot$mol^{-1}. (B) 10^{-3} cm$^3 \cdot$mol^{-1}.

(C) 1 cm$^3 \cdot$mol^{-1}. (D) 10^3 cm$^3 \cdot$mol^{-1}.

Knowledge Required: Terms in real gas equations of state and their approximate magnitude.

Thinking it Through: Two routes are possible. The first involves recognition of the magnitude from using the van der Waals equation numerically. The second involves recognizing that $V = \frac{4}{3}\pi r^3$ with $r \approx 10^{-8}$ cm$\cdot$molecule^{-1} or $V \approx (4 \times 10^{-24})(6 \times 10^{23}) \approx 24 \times 10^{-1} \approx 1$ cm$^3 \cdot$mol^{-1}; thus response **(C)** is correct.

ES-6. At 0.0°C and 200 torr, a pure gas sample has a density of 1.774 g·L^{-1}. The sample could consist of which of the following gases?

(A) N$_2$O (B) SF$_6$ (C) CO (D) C$_2$F$_4$

Equations of State

Knowledge Required: Relations within ideal gas equation of state.

Thinking it Through: The ideal gas equation of state is written as $PV = nRT$. We can substitute in that $n = \frac{m}{M}$ where M is the molar mass of the gas. We can rearrange the resulting equation to determine the molar mass from $M = \frac{m}{V}\frac{RT}{P}$. Substitution of the values in the problem gives

$$M = (1.774 \text{ g} \cdot \text{L}^{-1})\frac{(0.0821 \text{ L} \cdot \text{atm} \cdot \text{mol}^{-1} \cdot \text{K}^{-1})(273 \text{ K})}{(200/760) \text{ atm}} = 151 \text{ g} \cdot \text{mol}^{-1}.$$

The molar masses of the four gases are 44, 146, 28, and 100 respectively. Thus response **(B)** is correct.

ES-7. As pressure and temperature are increased to the critical point,

    (A) all of the following are true.

    (B) $\Delta_{\text{vaporization}}H$ goes to 0.

    (C) the density of the gas approaches the same value as that of the liquid.

    (D) the index of refraction of the gas approaches the same value as that of the liquid.

Knowledge Required: Critical point properties.

Thinking it Through: The critical point is the point at the highest temperature end of the liquid-vapor equilibrium, line in a one-component phase diagram. As pressure and temperature are increased along this line the properties of the vapor and the liquid move closer together until at pressures and temperatures above the critical point the material becomes a fluid (i.e., vapor and liquid are indistinguishable). Thus responses **(B)**, **(C)** and **(D)** are all correct, making response **(A)** the correct response.

ES-8. The valve between a 7.00 L tank containing Ne gas at 6.00 atm and a 5.00 L tank containing O_2 gas at 9.00 atm is opened. Calculate the final pressure in the tanks assuming ideal gas behavior and constant temperature.

    (A) 5.80 atm   **(B)** 7.25 atm   **(C)** 7.50 atm   **(D)** 7.75 atm

Knowledge Required: Calculations of final pressure for mixtures.

Thinking it Through: The final pressure is given by $P_{\text{final}} = 6.00 \text{ atm}\left(\frac{7.00}{12.00}\right) + 9.00 \text{ atm}\left(\frac{5.00}{12.00}\right) = 7.25$ atm or response **(B)** is correct.

ES-9. An equation of state for a non-ideal gas is $P(V - nb) = nRT$. The coefficient of thermal expansion, α, of any gas is defined as $\alpha = \left(\frac{1}{V}\right)\left(\frac{\partial V}{\partial T}\right)_P$. Therefore α for this gas is

    (A) $\frac{1}{T}$.   **(B)** $\frac{R}{T} + bP$.   **(C)** $nRT + nb$.   **(D)** $\frac{(R/P)}{(RT/P + b)}$.

Knowledge Required: Calculation of thermodynamic properties from a gas equation.

Thinking it Through: We need to perform the calculation of α from the equation of state as follows

$\left(\frac{\partial V}{\partial T}\right)_P = \left(\frac{\partial\left(\left(nRT/P\right)+nb\right)}{\partial T}\right)_P = \frac{nR}{P}$. Thus $\alpha = \left(\frac{1}{nRT/P + nb}\right)\left(\frac{nR}{P}\right)$. Cancelling the factors of n shows that response **(D)** is the correct response.

ES-10. Which is zero for an ideal gas?

(A) $\left(\frac{\partial U}{\partial T}\right)_V$ (B) $\left(\frac{\partial H}{\partial T}\right)_P$ (C) $\left(\frac{\partial U}{\partial V}\right)_T$ (D) $\left(\frac{\partial P}{\partial V}\right)_T$

Knowledge Required: Thermodynamic properties of an ideal gas.

Thinking it Through: The ideal gas has a number of unusual values of thermodynamic functions. Among these are that these properties depend only on temperature making responses **(A)** and **(B)** non-zero. Response **(D)** is also incorrect as this derivative equals $-\frac{nRT}{V^2}$. Response **(C)** is the correct response as the derivative is equal to zero.

Practice Questions

1. The van der Waals equation of state for a gas is
$\left(P + \frac{n^2 a}{V^2}\right)(V - nb) = nRT$

   Which statement is true?

   (A) The term nb makes allowance for attractive forces between molecules.
   (B) The term $\frac{n^2 a}{V^2}$ makes allowance for the volume of the molecules.
   (C) The constants a and b have the same values for all gases.
   (D) If P and V are expressed in atm and liters, respectively, the units of a and b are liters²·atm·mol⁻² and liter·mol⁻¹, respectively.

2. The constant b in the van der Waal's equation is linearly proportional to
   (A) the radius of one molecule.
   (B) the force of attraction between two molecules.
   (C) the excluded volume per mole.
   (D) the temperature.

3. A He gas thermometer at constant volume is designed so that the pressure was 642.7 torr at 32.38°C. What is the temperature of a system for which P = 784.9 torr?
   (A) 250.18 K   (B) 312.69 K
   (C) 340.75 K   (D) 373.13 K

4. A 10-L drum of ether at 18°C is open to the atmosphere (P = 760 torr) and contains 6 L of liquid. The top is sealed and the drum dented so that it now has an 8-L capacity. If the vapor pressure of ether at 18°C is 400 torr, what is the pressure inside the dented, sealed drum?
   (A) 950 torr   (B) 1120 torr
   (C) 1520 torr (D) 2320 torr

Equations of State

5. Given the relationship
$$\left(\frac{\partial U}{\partial V}\right)_T = T\left(\frac{\partial P}{\partial T}\right)_V - P$$
select the expression that is equal to $\left(\frac{\partial U}{\partial V}\right)_T$ for a gas that obeys the van der Waals equation of state,
$$P = \frac{nRT}{V-nb} - \frac{n^2 a}{V^2}$$

   (A) $-\dfrac{nRT}{V-nb}$ (B) $\dfrac{n^2 a}{V^2}$

   (C) $\dfrac{nRT}{V-nb}$ (D) $\dfrac{nRT}{V-nb} - \dfrac{n^2 a}{V^2}$

6. In the van der Waals equation of state
$$P = \frac{nRT}{V-nb} - \frac{n^2 a}{V^2}$$
the term, nb, will increase as the

   (A) molecular diameter increases.
   (B) intermolecular forces increases.
   (C) temperature increases.
   (D) pressure increases.

7. The isothermal compressibility
$$\kappa_T = -\left(\frac{1}{V}\right)\left(\frac{\partial V}{\partial P}\right)_T,$$
for the hard sphere equation of state
$$P(V-nb) = nRT$$
is given by

   (A) $-RT + b$ (B) $\dfrac{-RT}{P^2}$

   (C) $\dfrac{1}{T}$ (D) $\dfrac{\left(\dfrac{RT}{P^2}\right)}{\dfrac{RT}{P} + b}$

8. The derivative of the enthalpy with respect to P at constant T, $\left(\dfrac{\partial H}{\partial P}\right)_T$, for an ideal gas is

   (A) 0   (B) C_P (C) $\dfrac{\alpha}{\kappa}$ (D) μ_{JT}

9. The Joule-Thompson coefficient will predict whether

   (A) an ideal gas cools or heats on expansion.
   (B) how much energy is required to raise the temperature of a gas 1°C.
   (C) the entropy change for a gas on isothermal expansion.
   (D) a real gas heats or cools on pressure change.

10. The sign of the Joule-Thompson coefficient can be predicted from

    (A) the magnitude of the van der Waals a constant only.
    (B) the magnitude of the van der Waals b constant only.
    (C) the equation of state of the real gas.
    (D) the critical pressure, temperature and molar volume.

Answers to Study Questions

1. D
2. D
3. C
4. D
5. C
6. B
7. A
8. B
9. D
10. C

Answers to Practice Questions

1. D
2. C
3. D
4. B
5. B
6. A
7. D
8. A
9. D
10. C

Laws of Thermodynamics and State Functions

In the study of thermodynamics, it is often convenient to draw an imaginary box around the part of the universe where the phenomenon that you want to study is occurring. What is inside the box is called the **system**, what is outside the box is called the **surroundings**. The system and the surroundings together are taken to constitute the **universe**.

The first and second laws of thermodynamics are statements of experience. They summarize centuries of observations distilled from experiments concerning energy in the forms of heat and work, and the allowed conversions of one form of energy into the other. The first law can be stated in terms of an absolute energy: "The Energy of the Universe is Constant", or symbolically in terms of changes in energy: "$\Delta U = q + w$" where ΔU is the change in energy for a process; q is the heat transferred and w is the work performed during the process. Chemists use the convention assigning $w < 0$, if the system does work, and $w > 0$, if work is done on the system by the surroundings; and $q < 0$ if heat flows out of the system into the surroundings while $q > 0$ when heat flows into the system from the surroundings.

The second law has its roots in the limitations on the conversion of heat energy into work energy. It can be stated as "The entropy of the universe never decreases" or $\Delta S \geq 0$, where the equal sign is true for reversible processes and the > sign applies to irreversible processes. An operational definition of entropy is given by $dS = dq_{rev}/T$ where dq_{rev} is the differential element (infinitely small quantity) of heat transferred during a **reversible** process at temperature T.

The third law is best expressed in terms of Boltzmann's definition of entropy: $S = k \ln W$. For an infinitely large perfect crystal at 0 K, there is only one arrangement of energy, so $W = 1$ and $S(0K, \text{perfect crystal}) = 0$. It establishes a basis for the calculation of an absolute entropy.

State functions are properties that depend on the state of the system. Internal energy U, enthalpy H, entropy S, Gibbs (free) energy G and Helmholtz (free) energy A are the most commonly encountered state functions. Changes in state functions do not depend on the path taken to go from the initial state to the final state. Heat and work are not state functions. They, and quantities like the genertized heat capacity, C, that are derived from them, are not state functions; though C_P and C_V are state functions because then the path is defined. Their magnitudes change with the details of the process taken to get from one state to another.

Study Questions

LT-1. One mole of an ideal gas undergoes an isothermal expansion from 10.0 bar to 1.0 bar either

(1) reversibly,
(2) against a non-zero constant external pressure or
(3) freely against a vacuum.

The respective values of the work's magnitude obtained from these processes are

(A) $w_1 = w_2 = w_3$
(B) $w_1 < w_2 < w_3$
(C) $w_3 > w_1 > w_2$
(D) $w_1 > w_2 > w_3$

Knowledge Required: Expansion work is negative. The magnitude is the absolute value, the positive value of a negative quantity. PV-work is defined as $-\int P_{ext}dV$ where P_{ext} is the external pressure opposing the expansion.

Thinking It Through: The magnitude is related to the area under the $P_{ext} - V$ curve. The optimal (maximum amount of) work is obtained when the external pressure and the gas pressure are nearly balanced, as in a reversible process. When the external pressure is zero, as is the case when expansion takes place into a vacuum, then no work is done in the process. Therefore situation (2) corresponds to an intermediate in terms of the work produced. The ranking of the magnitudes is that given by response **(D)**. Response **(A)** would imply that work is a state function, because the three processes go between the same states. Work is not a state function so response **(A)** cannot be true. Response **(B)** would be true if the question did not ask for the magnitude. Remember that a negative number with a larger magnitude is smaller than one with a smaller magnitude, e.g., −2000 J is less than −1000 J. It is also helpful to think of a number line. A number to the left of another on the line is the smaller one. Response **(C)** fails to have the correct ordering and paths 1, 2, 3.

LT-2.	In an adiabatic expansion of an ideal gas, which of the following is *always* true?
	(A) The work done by the gas on the surroundings is equal to the increase in the internal energy of the gas.
	(B) The temperature of the gas will rise.
	(C) No work is done on the gas by the surroundings.
	(D) The work done by the gas on the surroundings is equal to the decrease in the internal energy of the gas.

Knowledge Required: Definition of the term *adiabatic*. The first law of thermodynamics. The special property of ideal gases that internal energy depends only on temperature, not on volume or pressure.

Thinking it Through: In an adiabatic process, $q = 0$. Therefore $\Delta U = w$. In an expansion, the system does work on the surroundings, so $w < 0$ and U will decrease. Therefore response **(D)** is correct. Response **(A)** is the opposite of response **(D)**, so it cannot be correct. Because $\Delta U = \int C_V \, dT$ for any process for an ideal gas, and because ΔU will be negative, the temperature of the system must go down. Therefore response **(B)** is incorrect. If the expansion has occurred into a vacuum, then no work would have been done ($P_{ext} = 0$) and response **(C)** would have been correct. The problem asks for a statement that is always true, and response **(C)** is only true for a special case.

LT-3.	If each CO molecule in a carbon monoxide crystal has equal probability of being situated on a lattice site with one of two orientations CO or OC, the value of $S°$ at 0 K will be nearest
	(A) 0 J·K^{-1}·mol^{-1} **(B)** 4.18 J·K^{-1}·mol^{-1}
	(C) 5.76 J·K^{-1}·mol^{-1} **(D)** 8.31 J·K^{-1}·mol^{-1}

Knowledge Required: The third law of thermodynamics. The Boltzmann definition of entropy, $S = R \ln W$ (for a mole of material).

Thinking it Through: This description suggests that the carbon monoxide crystal is disordered, therefore it is not a perfect crystal and $S°$ will not be 0 J·K^{-1}·mol^{-1} at 0 K so response **(A)** can be eliminated. The Boltzmann formula tells us that the contribution at 0 K will be determined by the number of orientations that each molecule of CO can assume in the crystal. We are told that there are two orientations, so $S°$ at 0 K should be approximately $R \ln(2)$ where R is the ideal gas constant, 8.31 J·K^{-1}·mol^{-1}. Because ln 2 is about 0.69, response **(B)** which is about half R is too small. The correct answer is response **(C)**.

Laws of Thermodynamics and State Functions

LT-4.	The enthalpy changes for combustion of monoclinic and rhombic sulfur are shown in the figure. From these values calculate $\Delta H°$ for the process S(rhombic) → S(monoclinic)		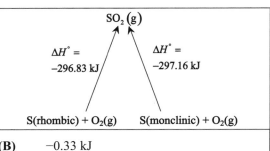
(A)	0.33 kJ	(B)	−0.33 kJ
(C)	−296.83 kJ	(D)	−593.99 kJ

Knowledge Required: Enthalpy is a State Function. Hess Cycles.

Thinking it Through: The two combustion steps can be combined to yield the desired reaction. In step II, the combustion step is reversed, and so the sign of the enthalpy change is reversed.

I. S(rhombic) + O_2(g) → SO_2(g) $\Delta H°$(I) = −296.83 kJ
II. SO_2(g) → S(monoclinic) + O_2(g) $\Delta H°$(II) = +297.16 kJ

Net: S(rhombic) → S(monoclinic) $\Delta H°$(Net) = $\Delta H°$(I) + $\Delta H°$(II)
                                     $\Delta H°$(Net) = +0.33 kJ

The correct response is **(A)**. If the direction of the process were reversed: monoclinic transforming to rhombic, the sign of $\Delta H°$ would be reversed, giving response **(B)**. Response **(C)** is just the combustion enthalpy of rhombic sulfur. Response **(D)** is the sum of the two enthalpies, if you forgot to change the sign of the enthalpy for Step II, this is the answer you would get.

LT-5.	For the adiabatic process illustrated in the figure, which is true?		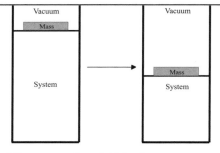
(A)	$q = 0$ and $\Delta U > 0$	(B)	$q = 0$ and $\Delta U < 0$
(C)	$q > 0$ and $\Delta U = 0$	(D)	$q < 0$ and $\Delta U = 0$

Knowledge Required: The First Law of Thermodynamics, definition of adiabatic process.

Thinking it Through: For an adiabatic process, $q = 0$ by definition. The figure shows the system decreasing in volume, thus w is positive by definition. Since $\Delta U = q + w = 0 + $ (positive) > 0. Thus response **(A)** is the correct response. Response **(B)** has an incorrect sign of ΔU. Responses **(C)** and **(D)** have $q \neq 0$ and are thus incorrect.

Laws of Thermodynamics and State Functions

LT-6. When a transformation occurs spontaneously at constant T and P, the signs of ΔG for the system and ΔS for the universe must be

	ΔG_{system}	$\Delta S_{universe}$
(A)	positive	positive
(B)	positive	negative
(C)	negative	positive
(D)	negative	negative

Knowledge Required: Thermodynamic functions for spontaneous reactions.

Thinking it Through: At constant T and P, ΔG of the system indicates the direction of a reaction with ΔG = negative, indicating the spontaneous direction. Thus responses **(A)** and **(B)** are incorrect. The second law of thermodynamics states that $\Delta S_{universe}$ must be positive for a spontaneous process, indicating that response **(C)** is correct.

LT-7. The molar heat capacity of diamond is adequately given by the equation $C_P/(J \cdot K^{-1} \cdot mol^{-1}) = 3.02 \times 10^{-7} \, T^3$. How much heat does it take to raise the temperature of one mole of diamond from 100 K to 300 K?

- **(A)** ½ $(3.02 \times 10^{-7})(300^2 - 100^2)$
- **(B)** ¼ $(3.02 \times 10^{-7})[300^4 - 100^4]$
- **(C)** $(3.02 \times 10^{-7})[1/(300) - 1/(100)]$
- **(D)** ⅓ $(3.02 \times 10^{-7})[300^3 - 100^3]$

Knowledge Required: Calculation of heat from heat capacity data.

Thinking it Through: The heat and heat capacity are related by $q = \int_{T_1}^{T_2} C_p \, dT$. For this particular problem, $q = \int_{100\,K}^{300\,K} (3.02 \times 10^{-7}) T^3 \, dT = (3.02 \times 10^{-7})(¼)(300^4 - 100^4)$. Thus response **(B)** is the correct response. The alternatives do not perform the integral correctly.

LT-8. Given the heat capacity information above, what is the absolute entropy of diamond at 300 K?

- **(A)** 0
- **(B)** ¼ $(3.02 \times 10^{-7})(300)^4$
- **(C)** $(3.02 \times 10^{-7})/(300)$
- **(D)** ⅓ $(3.02 \times 10^{-7})(300)^3$

Knowledge Required: Calculation of third law entropy from heat capacity data.

Thinking it Through: The third law entropy assuming the same heat capacity data from the previous problem is given by

$$S_{300} = S_0 + \int_0^{300\,K} \frac{C_P}{T} dT = 2.38 \, J \cdot mol^{-1} \cdot K^{-1} + (3.02 \times 10^{-7}) \int_0^{300\,K} \frac{T^3}{T} dT = 2.38 \, J \cdot mol^{-1} \cdot K^{-1} + (3.02 \times 10^{-7}) \frac{T^3}{3} \Big|_0^{300}$$

$= 2.38 \, J \cdot mol^{-1} \cdot K^{-1} + (⅓)(3.02 \times 10^{-7}) 300^3$. Thus response **(D)** is the correct response.

Laws of Thermodynamics and State Functions

LT-9. The equation relating the pressure-volume relationship for the reversible adiabatic expansion of an ideal gas is $P_1 V_1^\gamma = P_2 V_2^\gamma$. The equation relating the temperature-volume relationships for the same process is

(A) $\dfrac{T_2}{T_1} = \dfrac{V_1}{V_2}$

(B) $T_2 V_2^{\gamma-1} = T_1 V_1^{\gamma-1}$

(C) $\dfrac{T_1}{T_2} = \left(\dfrac{V_1}{V_2}\right)^{R/C_V}$

(D) $\dfrac{T_2}{T_1} = \left(\dfrac{V_1}{V_2}\right)^{C_P/C_V}$

Knowledge Required: P, V, T relationships for processes.

Thinking it Through: The reversible process has $P_1 V_1^\gamma = P_2 V_2^\gamma$. Substituting the ideal gas equation of state $P = nRT/V$ gives $\dfrac{nRT_1}{V_1} V_1^\gamma = \dfrac{nRT_2}{V_2} V_2^\gamma$ which simplifies to $\dfrac{T_1 V_1^\gamma}{V_1} = \dfrac{T_2 V_2^\gamma}{V_2}$ or $T_1 V_1^{\gamma-1} = T_2 V_2^{\gamma-1}$ and response **(B)** is the correct response.

LT-10. As the temperature approaches absolute zero, ΔG for any chemical reaction approaches

(A) ΔS. (B) ΔH. (C) T. (D) zero.

Knowledge Required: Definition of ΔG, limiting behavior with temperature.

Thinking it Through: The definition of $\Delta G = \Delta H - T\Delta S$. As T goes to 0, the second term goes to 0, thus ΔG goes to the value of the first term, namely ΔH. Thus response **(B)** is the correct response.

Practice Questions

1. $\Delta U°$ for the reaction

   $A(s) + 2B(g) \rightarrow 2C(s) + D(g)$

   was measured at temperature T in a constant volume calorimeter. $\Delta H°$ for the reaction is approximately

   (A) $\Delta U° - RT$
   (B) $\Delta U°$
   (C) $\Delta U° + RT$
   (D) $\Delta U° + 2RT$

2. An ideal gas undergoes irreversible isothermal expansion. Which is correct?

   (A) $\Delta S = -\Delta G / T$
   (B) $\Delta S = \Delta H / T$
   (C) $\Delta S = q / T$
   (D) $\Delta S = \Delta U / T$

3. The efficiency of a process with q_1 occurring at T_H and q_2 occurring at $T_C < T_H$ can be calculated from

   (A) $\dfrac{|w|}{q_1}$
   (B) $\dfrac{|w|}{|q_2|}$
   (C) $\dfrac{|q_2|}{w}$
   (D) $\dfrac{q_1}{|w|}$

4. A measure of the maximum non-PV work that can be performed by a process occurring at constant T and P is given by

   (A) ΔH.
   (B) ΔS.
   (C) ΔG.
   (D) ΔA.

5. A sample of water was placed in a refrigeration device and the temperature gradually lowered. Because there were no particles in the water to nucleate crystallization, the sample became supercooled and did not freeze until a temperature of –20°C was reached. When the water froze at –20°C, ΔS_{H_2O}, $\Delta S_{surroundings}$, and ΔS_{total} were

	ΔS_{H_2O}	$\Delta S_{surroundings}$	ΔS_{total}
(A)	−	−	−
(B)	+	+	+
(C)	0	0	0
(D)	−	+	+

6. The entropy change for a liquid heated from T_1 to T_2 can be calculated from the area under the curve obtained by plotting

(A) ΔH as the ordinate and $1/T$ as the abscissa.
(B) C_P as the ordinate and $\ln T$ as the abscissa.
(C) C_P as the ordinate and $1/T$ as the abscissa.
(D) C_P/T as the ordinate and $1/T$ as the abscissa.

7. In a cyclic process involving two steps, the first law of thermodynamics requires that

(A) $|q_1 + q_2| = |w_1|$
(B) $|q_1 + q_2| = |w_1 + w_2|$
(C) $|q_1 + q_2| > |w_1 + w_2|$
(D) $|q_1 + q_2| < |w_1 + w_2|$

8. From the information

$NH_3(g) \rightarrow N(g) + 3H(g)$ $\Delta H° = 1172.8$ kJ
$H_2(g) \rightarrow 2H(g)$ $\Delta H° = 435.930$ kJ
$N_2(g) \rightarrow 2N(g)$ $\Delta H° = 945.408$ kJ

the enthalpy of formation of $NH_3(g)$ is calculated to be

(A) $+2554.138$ kJ·mol^{-1}
(B) $+208.538$ kJ·mol^{-1}
(C) -46.2 kJ·mol^{-1}
(D) -1172.8 kJ·mol^{-1}

9. The third law of thermodynamics can be combined with experimental data to provide an absolute value for

(A) enthalpy. (B) work.
(C) entropy. (D) Gibbs energy.

10. Which is a graph of the enthalpy of vaporization from the triple point (T_p) to the critical point (T_c)?

(A)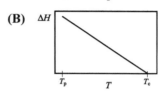

(B)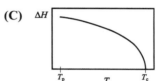

(C)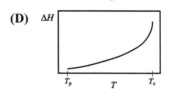

(D)

Answers to Study Questions

1. D
2. D
3. C
4. A
5. A
6. C
7. B
8. D
9. B
10. B

Answers to Practice Questions

1. A
2. A
3. A
4. C
5. D
6. B
7. B
8. C
9. C
10. C

Mathematical Relationships in Thermodynamics

While chemistry is an experimental science, it is generally not practical to make observations under all possible conditions. Thus, there are compelling reasons to build up an ability to manipulate the mathematical relationships that exist between the measurable variables. Even if a particular quantity is difficult or impossible to measure, as long as it can be related to more readily observed quantities, we can determine its value.

Thermodynamics is particularly suited to these types of mathematical formulation. Thermodynamic experiments are certainly carried out widely in scientific laboratories even today, but the fundamental tenets of this field and the mathematical machinery to describe them (at least for systems near equilibrium) are essentially set. This fact allows us to use the development of the mathematical relationships within thermodynamics as a template for understanding how mathematics can be applied to the physical sciences. From the more pragmatic perspective of the student, knowing the mathematical machinery and how it is derived means that even if a particular relationship is momentarily forgotten, we can always figure out what to do by falling back to the most fundamental concepts and equations and the mathematical rules for manipulating them.

From the perspective of the mathematics itself, the majority of the manipulations that need to be carried out in this type of test question involve the use of multi-variable differential calculus, particularly the manipulation of partial differentials and exact differentials. A review of this level of math is beyond the scope of this text, but there is a good chance that an appendix in the physical chemistry textbook you used in the course has reviews of this material.

Study Questions

MRT-1. For a pure substance the partial derivative of G with respect to P, $\left(\frac{\partial G}{\partial P}\right)_T$ is equal to,

(A) H.   (B) $-S$.   (C) T.   (D) V.

Knowledge Required: Expression for exact differential of Gibbs energy. Definition of partial derivative. The method for calculating partial derivatives.

Thinking it Through: For a closed one-component system in the absence of non-expansion work a change in G is proportional to a changes in P and T through:

$$dG = VdP - SdT$$

When T is constant the change in G is proportional only to a change in P:

$$dG = VdP$$

which through differentiation with respect to P leads to $\left(\frac{\partial G}{\partial P}\right)_T = V$. Thus response **(D)** is correct.

MRT-2. Which combination is **not** dependent on the path over which a thermodynamic process proceeds?

(A) $\Delta U, \Delta S, \Delta G, \Delta H$ (B) $\Delta U, q, \Delta S$

(C) $\Delta U, \Delta H, \Delta G, w$ (D) q, w

Mathematical Relationships in Thermodynamics

Knowledge Required: The meaning of state function.

Thinking it Through: Heat, q, and work, w, are not state functions and therefore are path dependent. Responses **(B)**, **(C)**, and **(D)** contain q, w, and q and w, respectively, and therefore contain path dependent functions so that they are incorrect. All functions listed in response **(A)** are state functions that do not depend on the path and therefore response **(A)** is correct.

MRT-3.	The total differential for H is		
	(A) $\quad dU$		**(B)** $\quad dU + VdP$
	(C) $\quad dU + VdP + PdV$		**(D)** $\quad dU - VdP - PdV$

Knowledge Required: Definition of enthalpy, H. The meaning of the term total differential.

Thinking it Through: By definition $H = U + PV$ meaning that H depends on U and PV where both, P and V are independent variables. The expression for the total differential dH gives the total change in H arising from changes in both U and PV. The change in U is represented by differential dU and the change in PV is represented by the differential $d(PV)$. Since both P and V are independent variables $d(PV)$ is equivalent to $VdP + PdV$. Response **(C)** contains all the required terms and therefore is correct. If you mistakenly define H as being equal to $U - PV$ you will obtain incorrect response **(D)**. Response **(A)** is a differential of U only and therefore is incorrect. In response **(B)** the term related to the volume change is missing and therefore this response is also incorrect.

MRT-4. Using the approximate equation of state for a gas, $V = \dfrac{nRT}{P} + nb$, ΔG associated with an isothermal change in pressure from P_1 to P_2 is

(A) $\quad nRT \ln\left(\dfrac{P_2}{P_1}\right) + nb(P_2 - P_1)$
(B) $\quad -nRT \ln\left(\dfrac{P_2}{P_1}\right)$

(C) $\quad -nRT \ln\left(\dfrac{P_2}{P_1}\right) + nb(P_2 - P_1)$
(D) $\quad nRT \ln\left(\dfrac{P_2}{P_1}\right)$

Knowledge Required: Calculation of thermodynamic functions associated with T and P changes, $\left(\dfrac{\partial G}{\partial P}\right)_T = V$ for all gases

Thinking it Through: $\Delta G_T = \int_{P_1}^{P_2} V\, dP = \int_{P_1}^{P_2}\left(\dfrac{nRT}{P} + nb\right) dP = nRT \ln \dfrac{P_2}{P_1} + nb(P_2 - P_1)$ Thus response **(A)** is the correct response.

MRT-5. If the effect of pressure on a reaction involving only pure solids is taken into account, then $\Delta G(T) = \Delta G^\circ(T) + \Delta V(P - P^\circ)$. Which statement is **not** relevant in deriving this equation?

(A) $\quad dG = -SdT + VdP$
(B) $\quad \left(\dfrac{\partial \Delta G}{\partial P}\right)_T = \Delta V$

(C) $\quad \dfrac{\partial^2 G}{\partial P \partial T} = \dfrac{\partial^2 G}{\partial T \partial P}$
(D) $\quad \Delta V$ is independent of pressure

Mathematical Relationships in Thermodynamics

Knowledge Required: Approximations made in calculating thermodynamic function changes on changes in T and P.

Thinking it Through: $\left(\dfrac{\partial \Delta G}{\partial P}\right)_T = \Delta V$ for all substances. This expression is then integrated from the standard state to arrive at the desired result assuming that ΔV is constant (independent of pressure). Thus responses **(B)** and **(D)** are relevant in deriving the equation. Response **(A)** is relevant in the derivation of $\left(\dfrac{\partial \Delta G}{\partial P}\right)_T = \Delta V$. Thus response **(C)** is not relevant in the derivation and is the correct response.

MRT-6. The differential for the Helmholtz function, A, at constant composition is $dA = -SdT - PdV$. The correct pair of equations that can be derived from this equation are

(A) $\left(\dfrac{\partial A}{\partial S}\right)_V = T$ and $\left(\dfrac{\partial A}{\partial V}\right)_V = -P$.

(B) $\left(\dfrac{\partial A}{\partial T}\right)_V = -S$ and $\left(\dfrac{\partial A}{\partial V}\right)_T = -P$.

(C) $\left(\dfrac{\partial A}{\partial S}\right)_P = T$ and $\left(\dfrac{\partial A}{\partial P}\right)_S = V$.

(D) $\left(\dfrac{\partial A}{\partial T}\right)_V = -S$ and $\left(\dfrac{\partial A}{\partial P}\right)_T = V$.

Knowledge Required: Derivation of Maxwell equations from the definitions of thermodynamic functions.

Thinking it Through: $\left(\dfrac{\partial A}{\partial T}\right)_V = -S$ and $\left(\dfrac{\partial A}{\partial V}\right)_T = -P$ after we take the derivative of A with respect to either T or V. For example, $\left(\dfrac{\partial A}{\partial T}\right)_V = \left(\dfrac{-S\partial T}{\partial T}\right)_V + \left(\dfrac{-P\partial V}{\partial T}\right)_V = -S(1) + (-P)0$. Thus response **(B)** is the correct response.

MRT-7. Select the partial derivative that is equal to the chemical potential of component i, μ_i

(A) $\mu_i = \left(\dfrac{\partial G}{\partial n_i}\right)_{S,V,n_k}$

(B) $\mu_i = \left(\dfrac{\partial G}{\partial n_i}\right)_{T,V,n_k}$

(C) $\mu_i = \left(\dfrac{\partial G}{\partial n_i}\right)_{T,H,n_k}$

(D) $\mu_i = \left(\dfrac{\partial G}{\partial n_i}\right)_{T,P,n_k}$

Knowledge Required: Definition of chemical potential as partial molar Gibbs energy at constant temperature and pressure.

Thinking it Through: All four responses have the differential part with the correct G and n_i. All four responses differ in what is held constant. The correct pair of functions is T, P (response **(D)** is correct) because $dG = -SdT + VdP + \sum \mu_i dn_i$.

MRT-8. The total differential for Gibbs free energy, G, is

(A) $dH - SdT$

(B) $dH - TdS$

(C) $dH - TdS - SdT$

(D) $dH + TdS + SdT$

Knowledge Required: Definition of G in terms of H, T, and S.

Thinking it Through: By definition $G = H - TS$. Taking the derivative gives $dG = dH - d(TS) = dH - TdS - SdT$ so that response **(C)** is the correct response.

Mathematical Relationships in Thermodynamics

MRT-9. A differential expression for the internal energy is $dU = TdS - PdV$. The corresponding Maxwell relation is

(A) $\left(\dfrac{\partial U}{\partial V}\right)_T = \left(\dfrac{\partial U}{\partial T}\right)_V$

(B) $\left(\dfrac{\partial S}{\partial P}\right)_T = \left(\dfrac{\partial V}{\partial T}\right)_S$

(C) $\left(\dfrac{\partial S}{\partial P}\right)_T = \left(\dfrac{\partial V}{\partial T}\right)_S$

(D) $\left(\dfrac{\partial T}{\partial V}\right)_S = -\left(\dfrac{\partial P}{\partial S}\right)_V$

Knowledge Required: Derivative of Maxwell relations.

Thinking it Through: The Maxwell relations are a recognition of the equality of the mixed second partials. Here the first partial derivatives are given by

$$T = \left(\dfrac{\partial U}{\partial S}\right)_V \text{ and } -P = \left(\dfrac{\partial U}{\partial V}\right)_S$$

The mixed second partials are then given by

$$\left(\dfrac{\partial T}{\partial V}\right)_S = \left(\dfrac{\partial}{\partial V}\left(\dfrac{\partial U}{\partial S}\right)_V\right)_S = \left(\dfrac{\partial}{\partial S}\left(\dfrac{\partial U}{\partial V}\right)_S\right)_V = \left(\dfrac{\partial}{\partial S}(-P)_S\right)_V$$

Eliminating the middle two terms shows that response **(D)** is the correct response.

MRT-10. For a rubber band $\left(\dfrac{\partial T}{\partial l}\right)_S = -\dfrac{T}{C_V}\left(\dfrac{\partial S}{\partial l}\right)_T$. The length of the rubber band

(A) increases with an increase in T.
(B) decreases with an increase in T.
(C) stays the same with an increase in T.
(D) can not be predicted with an increase in T.

Knowledge Required: The relationships between T, l, C_V, and S. Signs of ΔT, Δl, T, l, C_V, and S. Microscopic changes in rubber organization with T.

Thinking it Through: The temperature and heat capacity must be positive numbers. As the length of the rubber band increases (Δl is positive) the entropy decreases because the polymer is stretched and the number of possible configurations is reduced. Thus $\left(\dfrac{\partial S}{\partial l}\right)_T$ is negative, and the negative sign on the right-hand-side indicates that $\left(\dfrac{\partial T}{\partial l}\right)_S$ must be positive. The temperature of the rubber band will increase with length (an easy experiment to do, stretch a rubber band quickly while holding it against your lip; you should be able to feel the heat). Response **(A)** is the correct response.

Practice Questions

1. The Maxwell relation
$$\left(\dfrac{\partial S}{\partial V}\right)_T = \left(\dfrac{\partial P}{\partial T}\right)_V$$
follows from

(A) $TdS - PdV$
(B) $TdS + VdP$
(C) $-SdT + VdP$
(D) $-SdT - PdV$

2. Given the change in entropy, $\Delta S°$, and heat capacity, $\Delta C_P°$, for a reaction at 298 K, an estimate for $\Delta S°$ for the reaction at 50°C is

(A) $\Delta S°_{298} + \Delta C_P° (323 - 298)$

(B) $\Delta S°_{298} + \Delta C_P° \ln\left(\dfrac{50}{25}\right)$

(C) $\Delta S°_{298} + \Delta C_P° \ln\left(\dfrac{298}{323}\right)$

(D) $\Delta S°_{298} + \Delta C_P° \ln\left(\dfrac{323}{298}\right)$

Mathematical Relationships in Thermodynamics

3. Assuming that ΔH° is independent of temperature, the expression $\left(\dfrac{\partial \ln K_P}{\partial T}\right) = \dfrac{\Delta H^\circ}{RT^2}$ can be integrated to yield the expression:

(A) $\ln[K_P(T)] = \ln[K_P(T_0)] + \left(\dfrac{\Delta H^\circ}{R}\right)\left(\dfrac{1}{T_0^2} - \dfrac{1}{T^2}\right)$

(B) $\ln[K_P(T)] = \ln[K_P(T_0)] + \left(\dfrac{\Delta H^\circ}{R}\right)\left(\dfrac{1}{T^2} - \dfrac{1}{T_0^2}\right)$

(C) $\ln[K_P(T)] = \ln[K_P(T_0)] + \left(\dfrac{\Delta H^\circ}{R}\right)\left(\dfrac{1}{T} - \dfrac{1}{T_0}\right)$

(D) $\ln[K_P(T)] = \ln[K_P(T_0)] + \left(\dfrac{\Delta H^\circ}{R}\right)\left(\dfrac{1}{T_0} - \dfrac{1}{T}\right)$

4. The molar heat capacity of copper is adequately given by the equation
$C_P = 4.73 \times 10^{-5} T^3 \; (\text{J·mol}^{-1}\text{·K}^{-1})$
What is the absolute entropy of copper at 20 K?

(A) 2.4×10^{-6} J·mol^{-1}·K^{-1}
(B) 0.019 J·mol^{-1}·K^{-1}
(C) 0.13 J·mol^{-1}·K^{-1}
(D) 1.9 J·mol^{-1}·K^{-1}

5. Which partial derivative is always equal to zero for an ideal gas?

(A) $\left(\dfrac{\partial G}{\partial P}\right)_T$
(B) $\left(\dfrac{\partial H}{\partial T}\right)_P$
(C) $\left(\dfrac{\partial U}{\partial V}\right)_T$
(D) $\left(\dfrac{\partial V}{\partial T}\right)_P$

6. The relationship between the chemical potential of a substance and the partial pressure is

(A) $\mu_i = \mu_i^\circ + RT \ln(P_i / P^\circ)$
(B) $\mu_i = nRT / P_i$
(C) $\left(\dfrac{\partial \mu_i}{\partial V}\right)_T = P_i$
(D) $d\mu_i = -S_i dT - P_i dV$

7. In deriving $C_P - C_V$, which equation is used

(A) $dU = dq + dw$
(B) $H = U + PV$
(C) $\left(\dfrac{\partial U}{\partial V}\right)_T = 0$
(D) $PV = nRT$

8. The thermodynamic equation
$\left(\dfrac{\partial H}{\partial P}\right)_T = V - T\left(\dfrac{\partial V}{\partial T}\right)_P$ allows one to calculate the pressure dependence of molar enthalpy from P-V-T data. Consider a gas whose equation of state is $PV = 2nRT + nBP$, where B is a parameter which depends only on T. The pressure dependence of molar enthalpy for such a gas is

(A) $nR/P + \dfrac{dB}{dT}$.
(B) $nB - nT\dfrac{dB}{dT}$
(C) nB.
(D) zero.

9. Among the relationships derived from the fact that G is a state function is

(A) $\left(\dfrac{\partial S}{\partial T}\right)_P = \dfrac{C_P}{T}$
(B) $0 = \sum n_i d\mu_i$
(C) $\ln\left(\dfrac{K_{P,2}}{K_{P,1}}\right) = -\dfrac{\Delta H}{R}\left(\dfrac{1}{T_2} - \dfrac{1}{T_1}\right)$
(D) $G = H - TS$

10. Which expression shows the relationship between the chemical potentials of two components in one phase at equilibrium?

(A) $\mu_1 = \mu_2$
(B) $\mu_1 = (n_1/n_2)\mu_2$
(C) $d\mu_1 = -(n_2/n_1)d\mu_2$
(D) $\mu_1 = -(n_2/n_1)\mu_2$

Answers to Study Questions

1. D
2. A
3. C
4. A
5. C
6. B
7. D
8. C
9. D
10. A

Answers to Practice Questions

1. D
2. D
3. D
4. C
5. C
6. A
7. B
8. A
9. B
10. C

Chemical and Phase Equilibria

For pure materials, the chemical potential μ is equivalent to a molar Gibbs energy G_m. For materials in solutions, the chemical potential of the ith substance μ_i is given by $(\partial G/\partial n_i)_{T,P}$. It can be regarded as the Gibbs energy contributed per mole of substance to the solution. Just as balls roll downhill seeking a state with a lower gravitational potential and electrons move to a state of lower electrical potential, matter moves spontaneously to a state of lower chemical potential. The transformation of matter, both in phase changes and chemical processes, can be understood in terms of a driving force to lower chemical potential.

When multiple phases are in equilibrium, the chemical potential of each component is the same in every phase in which that component appears. Phase transformations occurring when the temperature is changed can be understood as the differing response of the chemical potential in the two phases to the temperature: $(\partial \mu_i/\partial T)_P = -S_i$ due to entropy differences in the phases. Pressure effects on phase equilibria arise because of differences in the molar volumes of the phases: $(\partial \mu_i/\partial P)_T = V_i$. Phase diagrams for one component systems (P vs T) represent the phase which is stable at a given P and T. Multiple component phase diagrams are usually represented as P vs composition at a fixed T or T vs composition at a fixed P. The Gibbs Phase Rule is given by $F = 2 + C - P$, where F is the number of variables that can be adjusted without requiring a phase change, 2 represents the variables T and P, C is the number of independent chemical components in the system and P is the number of phases present. The Phase Rule explains features observed in phase diagrams, such as the existence of triple points, but not quadruple points in one component systems, and horizontal lines in the phase diagrams representing three-phase equilibria.

The chemical potential for a substance under a general set of conditions of pressure and composition can be expressed in terms of the chemical potential of a standard state and the activity of the substance referred to that standard state: $\mu_i = \mu_i^\circ + RT \ln a_i$. For gases, the standard state is taken as the ideal gas at 1 bar pressure, and $a(\text{gas}) = P/1$ bar. For pure solids and liquids, the standard state is taken to be the pure material under an external pressure of 1 bar. Unless the problem involves very high pressures, the activities of pure condensed phases are taken to be unity. For components in solutions, there are two choices of standard states. One, called the Raoult's law or solvent standard state, is based on the behavior of the pure liquid; the other, called the Henry's law standard state is based on the limiting behavior of a solute at infinite dilution, in which a solute molecule is completely surrounded by solvent molecules.

The Gibbs free energy change ΔG for a chemical process, $a\text{A} + b\text{B} \rightleftharpoons c\text{C} + d\text{D}$, can be expressed as

$$\Delta G = \Delta G^\circ + RT \ln \left(\frac{a_C^c a_D^d}{a_A^a a_B^b} \right)$$

where ΔG° is the free energy change with reactants and products in their standard states and the ln term contains the activities of each substance under the current experimental conditions, relative to the standard state of each term. The ratio of activities is sometimes referred to as the reaction quotient, and designated Q. For a spontaneous process $\Delta G < 0$. At equilibrium, $\Delta G = 0$ and

$$\Delta G^\circ = -RT \ln \left(\frac{a_C^c a_D^d}{a_A^a a_B^b} \right)_{eq}$$

The ratio of activities at equilibrium is given a special name and symbol, the equilibrium constant, K.

The activities of electrolyte solutions are treated in terms of mean ionic activity coefficients because individual ion effects cannot be separated out. The hypothetical 1 molal Henry's law solution is taken as the standard state of an ionic solution. At low concentrations, the mean ionic activity coefficient of an ionic solute in an aqueous solution at 298.15 K can be estimated from the Debye-Hückel limiting law:

$$\log \gamma_\pm = -0.509 |z^+ z^-| I^{1/2}$$

where $\gamma_\pm$ is the mean ionic activity coefficient of the solute with ionic charges z^+ and z^- in a solution of ionic strength I.

The electrical work performed during the operation of an electrochemical cell can be related to the ΔG of the process taking place within the cell: $\Delta G = -nFE$ where n is the number of moles of electrons transferred, F is the Faraday constant and E is the cell potential. The Nernst equation shows how the activities of reactants and products affect the cell potential:

$$E = E^\circ - \frac{RT}{nF} \ln\left(\frac{a_C^c a_D^d}{a_A^a a_B^b}\right)$$

where A and B are reactants in the electrochemical process and C and D are products.

Study Questions

CPE-1. At the two points, **A** and **B**, on the solid-gas phase transition line on the phase diagram,

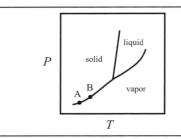

- **(A)** the μ's of the solids are equal.
- **(B)** the μ's of the gases are equal.
- **(C)** the μ of the solid is equal to the μ of the gas at each point.
- **(D)** all the μ's are equal everywhere the two phases coexist.

Knowledge Required: Lines on a $P - T$ phase diagram represent two-phase equilibria; chemical potentials are equal when phases are in equilibrium, Gibbs energy.

Thinking it Through: Both points lie on the same two phase line. At each point on a two phase line, the chemical potential of one phase is equal to that of the other: $\mu_I = \mu_{II}$ where I and II represent two different phases, therefore response **(C)** is the correct one. Points A and B are at different temperatures and pressures, because chemical potentials change with both T and P, $(\partial \mu_i /\partial T)_P = -S_i$ and $(\partial \mu_i /\partial P)_T = V_i$, it is unlikely that the chemical potentials of either the gases or the solids would stay the same for an arbitrary change in both T and P, so responses **(A)** and **(B)** would not be expected to be true in general. Response **(D)** also ignores the temperature and pressure dependences of μ.

CPE-2. At 25°C and $P^\circ = 1$ bar, ΔG°, the Gibbs energy change for the transformation

$$\text{graphite} \rightleftharpoons \text{diamond}$$

is positive. The volume change $\Delta V = V_d - V_g$ is negative. An approximate expression for the pressure at which diamond will be in equilibrium with graphite at 25° C is

- **(A)** $P^\circ - (\Delta G^\circ/\Delta V)$
- **(B)** $P^\circ + (\Delta G^\circ/\Delta V)$
- **(C)** P°
- **(D)** $\Delta G^\circ/\Delta V$

Knowledge Required: Conditions for equilibrium: chemical potentials of two phases are equal and so $\Delta G = 0$. Change in chemical potentials (and Gibbs energy change) with pressure: $(\partial \mu / \partial P)_T = V$ and $(\partial \Delta G / \partial P)_T = \Delta V$.

Thinking It Through: We can write a total differential for ΔG as a function of T and P:

$$d\Delta G = \Delta V dP - \Delta S dT$$

but since the process is isothermal, we need only consider the dP term. If we integrate both sides of the equation for the pressure change between $P°$ and the equilibrium pressure we obtain,

$$\int_{\Delta G°}^{\Delta G_{eq}} d\Delta G = \int_{P°}^{P_{eq}} \Delta V \, dP$$

and remember that $\Delta G = 0$ at the new pressure we get

$$-\Delta G° = \Delta V (P_{eq} - P°)$$

Rearranging and solving for the equilibrium pressure gives us

$$P_{eq} = -\frac{\Delta G}{\Delta V} + P°$$

The correct response is response **(A)**.

CPE-3. Benzene and ethanol form an azeotropic mixture that is 60 mol% benzene and has a boiling point of 341 K at 1 atm. The normal boiling points of pure benzene and pure ethanol are 353 K and 352 K, respectively. Which statement best describes the behavior of benzene-ethanol solutions.

(A) Solutions of any composition of the two materials can be completely separated into the two components by a sufficiently efficient distillation column.

(B) The vapor phase in equilibrium with a benzene-ethanol solution of any composition is richer in ethanol than is the liquid phase.

(C) The solutions show positive deviations from Raoult's Law.

(D) The solutions show negative deviations from Raoult's Law.

Knowledge Required: Vapor-liquid equilibria; meaning of azeotrope. Raoult's Law and deviations of Raoult's law.

Thinking it Through: First, it is helpful to have in mind a qualitative picture of the temperature (T) – composition (x) phase diagram for this system. A rough sketch is given on the left below. The diagram shows that the lowest boiling point mixture in the $T-x$ phase diagram will be the azeotrope. Whenever there is an azeotrope, complete separation of the two components is not possible, therefore response **(A)** cannot be correct. The two vertical lines in the figure show two possible liquid compositions, one on either side of the azeotrope. The arrows show that the vapor composition formed by each liquid approaches that of the azeotrope – for the liquid on the ethanol-rich side, the vapor is richer in benzene, for the liquid on the benzene-rich side, the vapor is richer in ethanol. Therefore, response **(B)** cannot be correct. To consider responses **(C)** and **(D)**, it is helpful to think of the pressure - composition diagram that would complement this one. Again, a qualitative sketch is all that is needed. The $P-x$ diagram can be constructed by remembering that a high temperature boiling liquid corresponds to one with a low vapor pressure while a liquid with a low boiling point has a high vapor pressure. Applying that rule to the benzene-ethanol system, which has a minimum-boiling azeotrope, shows that the azeotrope will have a higher vapor pressure than either pure benzene or pure ethanol. The dashed line in the $P-x$ diagram represents the vapor pressures that would be found above ideal solutions. The arrows indicate that the vapor pressures of the real solutions are above those of the ideal solutions (ones which follow Raoult's law). Because the vapor pressures are greater than those predicted by Raoult's law, we have a positive deviation. Thus, response **(C)** is the correct one. An azeotrope with a minimum in the vapor pressure curve would deviate negatively from Raoult's law behavior which is response **(D)**.

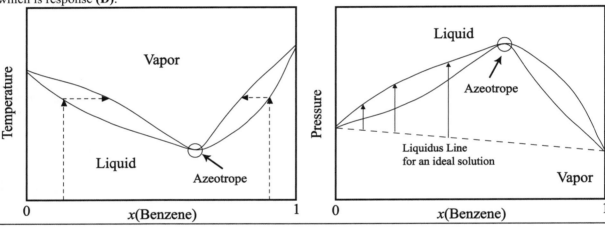

CPE-4. Choose the statement that describes the relationship between the solubility of AgCl in 0.10 M KNO_3 and in pure water.

    (A) AgCl is more soluble in 0.10 M KNO_3 because the thermodynamic equilibrium constant is greater in 0.10 M KNO_3 than in pure water.

    (B) AgCl is more soluble in 0.10 M KNO_3 because the activity coefficients of Ag^+ and Cl^- are smaller in 0.10 M KNO_3 than in pure water.

    (C) AgCl is more soluble in 0.10 M KNO_3 because the activity coefficients of Ag^+ and Cl^- are greater in 0.10 M KNO_3 than in pure water.

    (D) AgCl is equally soluble in 0.10 M KNO_3 and in pure water because AgCl and KNO_3 have no ions in common.

Knowledge Required: Solubility Constants; Mean Ionic Activity Coefficients and Debye-Hückel Activity Coefficient Relationships.

Thinking it Through: The reaction of interest is $AgCl(s) \rightleftharpoons Ag^+(aq) + Cl^-(aq)$ with the equilibrium constant given by

$$K = m_{Ag^+} m_{Cl^-} \gamma_\pm^2$$

While it is true that AgCl and KNO$_3$ have no ions in common, the mean ionic activity coefficients are affected by the total ionic strength of the solution, I. I is a sum over all the ions in solution, so the presence of K$^+$ and NO$_3^-$ ions increases I. Both the limiting form of the Debye-Hückel relationship

$$\log \gamma_\pm = -A \, |z^+ z^-| \, I^{1/2}$$

and the extended Debye-Hückel equation

$$\log \gamma_\pm = -\frac{A \, |z^+ z^-| \, I^{1/2}}{1 + I^{1/2}}$$

predict that the mean ionic activity coefficient will decrease with increasing ionic strength. (As I increases, the right hand side becomes more negative. A negative number implies an activity coefficient which is less than unity.) As $\gamma_\pm$ decreases, the molalities of Ag$^+$ and Cl$^-$ must increase to maintain the equality of the right hand side with the solubility product. Therefore the solubility will be increased. The correct response is **(B)**.

CPE-5. Given the potentials at 25°C for the following half cells:

	$E°/V$
$O_2(g) + 2 H_2O(l) + 4 e^- \rightarrow 4 OH^-(aq)$	0.401
$O_3(g) + H_2O(l) + 2 e^- \rightarrow O_2(g) + 2 OH^-(aq)$	1.240

what is $\ln K_P$ at 25°C for the reaction $2 O_3(g) \rightarrow 3 O_2(g)$?

(A) $\dfrac{4F(-2.079V)}{RT}$ (B) $\dfrac{2F(-0.839V)}{RT}$

(C) $\dfrac{4F(+0.839V)}{RT}$ (D) $\dfrac{4F(+2.079V)}{RT}$

Knowledge Required: Half cell reactions and cell potentials; relationship of ΔG and E

Thinking It Through: We can get $\ln K_P$ from $-nFE° = \Delta G° = -RT \ln K_P$. We get $\Delta G°$ for the reaction by combining the half-cell potentials appropriately. Because O$_2$ gas is the product in our desired reaction, the first half-cell must be reversed. Then, in order to cancel the electrons, we must multiply the ozone half-cell by two but remember not to multiply the voltage.

		$E°/V$
	$2[O_3(g) + H_2O(l) + 2 e^- \rightarrow O_2(g) + 2 OH^-(aq)]$	1.240
	$4 OH^-(aq) \rightarrow O_2(g) + 2 H_2O(l) + 4 e^-$	-0.401
Net	$2 O_3(g) \rightarrow 3 O_2(g)$	$\Delta G° = -nFE° = -4F(0.839V) = -323$ kJ

Remember that half-cell potentials are intensive quantities independent of the number of electrons transferred, but $\Delta G°$ is extensive. The correct answer is response **(C)**. Responses **(A)** and **(D)** come from multiplying the ozone half-cell potential by two. Response **(B)** has the wrong number of electrons for the half cell.

Chemical and Phase Equilibria

CPE-6. The thermodynamic solubility product constant of silver acetate is 4.0×10^{-3}. The solubility of silver acetate in a sodium nitrate solution is 0.094 mol·L^{-1}. The mean ionic activity coefficient of silver acetate in this solution is

(A) 0.063 (B) 0.188 (C) 0.67 (D) 1.49

Knowledge Required: Definition of thermodynamic solubility product constant. Relationship among solubility product constant, solubility, and ionic activity coefficient.

Thinking It Through: The definition of the thermodynamic solubility product constant in terms of the ionic activity coefficients and solubilities is

$$K_{sp} = \gamma_{Ag^+}[Ag^+]\gamma_{Ac^-}[Ac^-]$$

On substituting the data in the problem and rearranging slightly

$$\gamma_{Ag^+}\gamma_{Ac^-} = \frac{4.0 \times 10^{-3}}{(0.094)^2} = 0.45$$

and the mean ionic activity coefficient is the square root of this number which is 0.67. Thus the correct response is then response **(C)**.

CPE-7. A 6.0 mol sample of benzene is mixed adiabatically with 2.0 mol of toluene at 25°C. Assume this is an ideal solution. What is the entropy for the mixing process?

(A) zero
(B) $-R\,[0.75\ln(0.75) + 0.25\ln(0.25)]$
(C) $-R\,[6.0\ln(0.75) + 2.0\ln(0.25)]$
(D) $R\,[6.0\ln(0.75) + 2.0\ln(0.25)]$

Knowledge Required: Thermodynamic functions for mixing ideal solutions.

Thinking It Through: The expression for the entropy of mixing is $\Delta S = -R\sum n_i \ln x_i$. For this mixture $n_{tot} = 8.0$ mol, $x_{benzene} = \frac{6.0}{6.0+2.0} = 0.75$ and $x_{toluene} = \frac{2.0}{6.0+2.0} = 0.25$. Substituting in the values gives $\Delta S = -R[6.0\ln(0.75) + 2.0\ln(0.25)]$, so that response **(C)** is the correct response. Response **(A)** gives no mixing, response **(B)** is similar but missing the n_i, it uses x_i of each component instead. Response **(D)** is missing the correct sign.

CPE-8. Two phases are considered to be in mutual equilibrium when

(A) both are at their critical temperatures.
(B) the mole fractions of all components are the same in each phase.
(C) the temperature, pressure, and chemical potential of each component are the same in each phase.
(D) molecules are moving across the phase boundary.

Knowledge Required: Conditions for equilibrium in multi-component systems.

Thinking It Through: Equilibrium is reached when the change in Gibbs energy, ΔG, is 0. ΔG depends on the temperature, pressure, and chemical potential of each species present. Response **(A)** illustrates a "special" case of equilibrium which doesn't occur except for a single point. Response **(B)** suggests equilibrium occurs when the same amount of each component is in each phase (moles balance), which is not correct. Response **(D)** is the nature of dynamic equilibrium on the molecular level, response **(C)** is the correct response.

Chemical and Phase Equilibria

CPE-9. Which statement about the reaction $H_2(g) + D_2(g) \rightleftharpoons 2HD(g)$ is correct?

(A) The driving force for the reaction is the enthalpy of reaction.

(B) The driving force for the reaction is the entropy of the reaction.

(C) The driving force for the reaction is the difference in bond energies between products and reactants.

(D) The change in the standard Gibbs energy for the reaction is zero.

Knowledge Required: Driving forces for equilibrium, isotopic exchange reaction.

Thinking It Through: In order for a reaction to occur there must be a driving force corresponding to a change in ΔG for this process is given by $\Delta G = \Delta H - T\Delta S$. An isotopic exchange reaction such as that in the problem will be approximately thermoneutral, i.e. $\Delta H \approx 0$. Thus response (A) must be incorrect. Response (C) is also an answer based on bond energies or enthalpy which has been shown to be incorrect. Response (B) is correct since the entropy of the reaction must increase with the increased arrangement disorder of the products. Response (D) would require that both ΔH and ΔS be zero, which has just been shown to be incorrect.

CPE-10. For the conversion of A to B at 25°C and 1 atm, $\Delta H° = 1.90$ kJ·mol^{-1}. The density of B is greater than that of A at 25°C and 1 atm. The condition of temperature and pressure which favor the formation of product B over reactant A are

(A) high temperature, high pressure. (B) high temperature, low pressure.

(C) low temperature, high pressure. (D) low temperature, low pressure.

Knowledge Required: Dependence at equilibrium on temperature and pressure.

Thinking It Through: The information given in the problem is that the reaction is endothermic, which implies that product B will be favored under high temperature conditions. In addition, the information is that the density of B is greater than the density of A. Thus the molar volume of the product is less than that of the reactant. This implies that product will be favored at higher pressure. The correct response is the combination of high temperature and pressure; response (A) is the correct response.

Practice Questions

1. For many substances near the melting point of the solid

      density of liquid < density of solid

   and the Clapeyron equation $(dP/dT) = (\Delta S /\Delta V)$ can be used to estimate the effects of pressure on the melting point temperature. Increasing the pressure on such a system by a factor of two will cause the melting point temperature to

   (A) decrease.

   (B) increase.

   (C) remain the same.

   (D) always increase by a factor of 2.

2. The vapor pressure of pure water at 25°C is 23.76 torr. Which value represents the vapor pressure of an aqueous solution of a nonvolatile, weak monoprotic acid at a concentration of $x_{HA} = 0.010$.

   (A) 0.238 torr   (B) 23.38 torr

   (C) 23.52 torr   (D) 23.76 torr

Chemical and Phase Equilibria

3. The $\Delta_f G°$ values for *n*-butane and *i*-butane at 25°C are

	$\Delta_f G°$ / kJ·mol^{-1}
n-butane	−15.71
iso-butane	−17.97

   When an equimolar mixture of *n*-butane and *iso*-butane is allowed to attain an equilibrium state between the two isomers

   (A) no *iso*-butane remains.
   (B) more *iso*-butane than *n*-butane is present.
   (C) the pressure of *n*-butane is zero.
   (D) the exact amounts of the two isomers cannot be determined from this information.

4. The standard Gibbs energy change, $\Delta G°$, for a certain chemical reaction is −10.0 kJ·mol^{-1}. The equilibrium constant at 300 K is about

   (A) 0.0181
   (B) 1.00
   (C) 4.20
   (D) 55

5. When a transformation occurs at a constant volume and temperature, the maximum work which can appear in the surroundings is equal to

   (A) $-\Delta A$.
   (B) $-\Delta G$.
   (C) $-\Delta H$.
   (D) $-\Delta S$.

6. If a chemical system is at equilibrium at constant *T* and *P*, what thermodynamic function for the system must have a minimum value?

   (A) C_P
   (B) H
   (C) G
   (D) S

7. When a transformation occurs spontaneously at constant *T* and *V*, the signs of ΔA for the system and ΔS for the universe must be

	$\Delta A_{(system)}$	$\Delta S_{(universe)}$
(A)	positive	positive
(B)	positive	negative
(C)	negative	positive
(D)	negative	negative

8. Select the thermodynamic criterion for a spontaneous process in a closed **constant pressure** system immersed in a constant temperature bath.

   (A) $\Delta U < 0$
   (B) $\Delta G < 0$
   (C) $\Delta A < 0$
   (D) $\Delta S > 0$

9. Select the thermodynamic criterion for a spontaneous process in a **constant volume** system immersed in a constant temperature bath.

   (A) $\Delta U < 0$
   (B) $\Delta G < 0$
   (C) $\Delta A < 0$
   (D) $\Delta S > 0$

10. When the reaction, $A(g) \rightleftharpoons B(g)$, reaches equilibrium at a constant total pressure of 1 atm and a constant temperature, *T*, the pressure of A is twice that of B $(P_A = 2P_B)$. What is the value of $\Delta G°$ for this reaction?

    (A) $-RT \ln 6$
    (B) $RT \ln 2$
    (C) $RT \ln 3$
    (D) $RT \ln 6$

Answers to Study Questions

1. C
2. A
3. C
4. B
5. C
6. C
7. C
8. C
9. B
10. A

Answers to Practice Questions

1. B
2. C
3. B
4. D
5. A
6. C
7. C
8. B
9. C
10. B

Kinetic Molecular Theory

Physical chemistry deals with many gaseous systems, in large part because they can be described using a relatively uncomplicated model for bulk behavior in terms of the actions of atoms or molecules. Thus, while applications including solids and liquids are important to think about in terms of molecular motions, the gas phase is normally the starting point for developing models capable of building this bridge. The kinetic molecular theory of gases is used to describe the bulk behavior of gases based on molecular motions.

Because kinetic molecular theory addresses the behavior of molecules, which are individually invisible, it must begin with postulates about the system. To develop the model, we assume the postulates to be true, though they may address factors that are impossible to observe, and then infer what macroscopic samples of the gases must do based on these postulates. Thus, kinetic molecular theory begins by postulating that

- Gases are composed of particles in constant random motion.
- The volume of space between particles in a gas is much larger than the volume of the particles themselves, so their volumes can be ignored.
- Particles in a gas do not interact with each other or the walls of their container except when they collide and the collisions are elastic when they occur.
- The temperature of a gas is related to the average kinetic energy of the particles in the gas.

It may seem difficult to imagine the wealth of information that can be derived from these few postulates, but a great deal about gases can be inferred from just this much information. The actual derivation of the facts of kinetic molecular theory is beyond the scope of this review, but we can look at questions that can be asked about them. Thus, in this chapter, the kinetic molecular theory is examined to yield information about the distribution of molecular velocities, and the number of collisions of molecules with other species and with the walls of the container.

Study Questions

KMT-1. An instantaneous "snapshot" of a sample of gaseous helium (bp = 4.2 K) at 298 K might look like the sketch at right.

Which snapshot best represents the sample after cooling helium to 250 K under constant volume conditions?

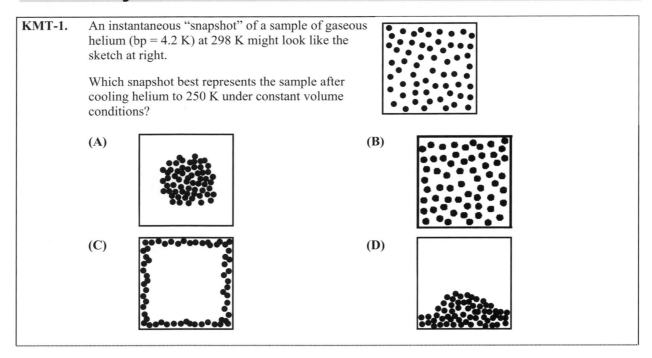

(A) (B) (C) (D)

Knowledge Required: Conceptual picture of arrangement of gaseous species.

Thinking it Through: The fundamental assumptions involved in the kinetic molecular theory of gases require that the gas molecules be evenly distributed throughout the container. The only response which has the molecules evenly distributed in response **(B)**, which must then be correct.

KMT-2.	The speed distribution function, $F(s)$, of a given gas at four temperatures is shown. Which trace corresponds to the highest temperature?			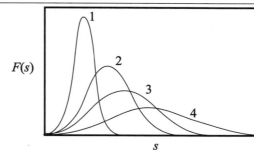			
	(A) 1	**(B)**	2	**(C)** 3	**(D)**	4	

Knowledge Required: Effects on distribution functions as mass and temperature change.

Thinking it Through: The speed distribution for a gas extends from 0 to $+\infty$ and looks like (constant) $\times s^2 \times e^{-ms^2/2k_BT}$. Responses **(B)** and **(C)** are intermediate temperatures and thus not reasonable answers. Trace 1 corresponds to a lower temperature than trace 4. In this problem, with constant mass, trace 4 corresponds to the higher temperature, and thus **(D)** is the correct response.

KMT-3.	For the hypothetical molecular speed distribution shown, the relationship between the most probable speed s_{mp}, the average speed s_{av} and the root-mean-square speed s_{rms} is		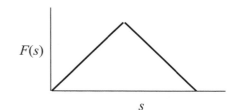	
(A) $s_{mp} < s_{av} < s_{rms}$.	**(B)** $s_{mp} < s_{rms} < s_{av}$.	**(C)** $s_{mp} = s_{av} < s_{rms}$.	**(D)** $s_{mp} = s_{av} = s_{rms}$.	

Knowledge Required: Relationship of most probable, average, and root mean square speeds to shape of probability distribution.

Thinking it Through: The highest point corresponds to the most probable speed. The symmetry of the function about the highest point makes the average speed and the most probable speed to both have the value at the highest point. The straight line behavior of the function causes the root mean square speed to also have the value at the maximum. Thus response **(D)** gives the correct response to the question.

KMT-4. The Maxwell distribution for molecular speeds can be written in the form:

$$F(s) = 4\pi \left(\frac{m}{2\pi k_B T}\right)^{3/2} s^2 e^{-(ms^2/2k_B T)} ds$$

in which s is the speed of the molecule. The other symbols have their standard meanings. For the total number of molecules, N, the average speed of a molecule can be calculated from the expression:

(A) $4\pi \left(\dfrac{m}{2\pi k_B T}\right)^{3/2} \displaystyle\int_0^\infty s^3 e^{-(ms^2/2k_B T)} ds$

(B) $4\pi N \left(\dfrac{m}{2\pi k_B T}\right)^{3/2} \displaystyle\int_0^\infty s^2 e^{-(ms^2/2k_B T)} ds$

(C) $4\pi N \left(\dfrac{m}{2\pi k_B T}\right)^{3/2} \displaystyle\int_0^\infty s^3 e^{-(ms^2/2k_B T)} ds$

(D) $2\pi m \left(\dfrac{m}{2\pi k_B T}\right)^{3/2} \displaystyle\int_0^\infty s^4 e^{-(ms^2/2k_B T)} ds$

Knowledge Required: Evaluation of averages from a probability distribution.

Thinking it Through: The average of a probability distribution is found by evaluating $\int F(s)\, s\, ds$ over the range of s. In this case, the speed the range extends from 0 to $+\infty$. Thus, the average speed is given by response **(A)**. Response **(B)** is incorrect for two reasons, there is an additional factor N in front of the integral, and s is raised to the second power rather than $s \cdot s^2 = s^3$. Response **(C)** is incorrect because of the factor of N in front of the integral. Response **(D)** is incorrect because of the factor of m in front of the integral and of s^4 within the integrand.

KMT-5. At what temperature will the average velocity of He be equal to that of N_2 at 298K?

(A) $(298\ K)(4/28)$

(B) $(298\ K)(28/4)^{-1/2}$

(C) $(298\ K)(4/28)^{-1/2}$

(D) $(298\ K)(28/4)$

Knowledge Required: Formula for the most probable, average, and root mean square velocities as function of mass and temperature.

Thinking it Through: The average velocity for a molecule at a given temperature is given by $\bar{v} = \left(\dfrac{8RT}{\pi M}\right)^{1/2}$, where R is the gas constant, T the temperature in K, and M the molar mass in kg·mol^{-1}. Setting the two average velocities equal we arrive at $\left(\dfrac{8RT_{He}}{\pi M_{He}}\right)^{1/2} = \left(\dfrac{8RT_{N_2}}{\pi M_{N_2}}\right)^{1/2}$ or on cancellation of terms

$T_{He} = T_{N_2}\left(\dfrac{M_{He}}{M_{N_2}}\right) = 298\left(\dfrac{4}{28}\right)$ or response **(A)** because responses **(B)** and **(C)** still have a square root (−½ or +½), and response **(D)** incorrectly has 28/4.

KMT-6. For O_2 at 300 K, which quantity is the largest?

(A) the average speed

(B) the most probable speed

(C) the root mean square speed

(D) all of the above have the same value

Knowledge Required: Formula for most probable, average, and root mean square speeds as function of mass and temperature.

Thinking it Through: The most probable, average, and root mean square speeds are related by $s_{mp} = \left(\dfrac{2RT}{M}\right)^{1/2}$, $s_{ave} = \left(\dfrac{8RT}{\pi M}\right)^{1/2}$, and $s_{rms} = \left(\dfrac{3RT}{M}\right)^{1/2}$. Because $3 > 8/\pi > 2$, the root mean square speed has the largest value of the three. Response **(C)** is thus the correct response.

KMT-7. For an ideal gas, the kinetic theory of gases predicts the number of collisions per unit time for any particular molecule to be directly proportional to

(A) pressure. (B) temperature.
(C) molar mass. (D) molecular diameter.

Knowledge Required: Definition of collision rates in terms of quantities, conversion of number densities.

Thinking it Through: The kinetic molecular theory of gases predicts that the number of collisions of one gaseous molecule is proportional to the number density of the other molecules in the gas phase. This proportionality to the number density makes the number of collisions proportional to the pressure of the gas phase molecules (response **(A)** is correct). The dependence on temperature is not a direct proportionality because it is inversely proportional to the temperature in the number density term and proportional to the square root of the temperature in the velocity term. The molar mass proportionality is also not direct but rather through the square root in the velocity term. The number of collisions is also proportional to the molecular diameter squared. Thus responses **(B)**, **(C)**, and **(D)** are incorrect because of the lack of direct proportionality; they are proportional to these properties to some power other than one.

KMT-8. At room temperature and pressure, what is the approximate collision frequency of nitrogen molecules with a 1−cm² surface?

(A) 10^{-10} s^{-1} (B) 10^{3} s^{-1} (C) 10^{23} s^{-1} (D) 10^{43} s^{-1}

Knowledge Required: Order of magnitude of various quantities related to collisions with other molecules and walls.

Thinking it Through: The formula for the number of collisions of gas phase molecules per unit time per unit area is $z_{wall} = (1/4)(N/V)\bar{s}$ where z_{wall} is the number of collisions per unit time per unit area, (N/V) is the number density of gas phase molecules, and $\bar{s}$ is the average speed. Since (N/V) is about 10^{19} molecule cm^{-3} for room temperature and pressure, and the average speed is about 10^{5} cm s^{-1}, z_{wall} is about 10^{23} s^{-1} (response **(C)**).

KMT-9. In a 1 L bulb at 1 bar, which noble gas is expected to have the largest hard sphere collision cross section at 300 K?

(A) Ne (B) Ar (C) Kr (D) Xe

Knowledge Required: Relative size of molecules and how that affects collisional properties.

Thinking it Through: In general, the molecular size (radius, and thus diameter) increase as you descend through the periodic table within a given group. Thus you would expect Xe to be the largest of this group of Group 18 elements, and response **(D)** would be correct.

Kinetic Molecular Theory

KMT-10. In the gas phase, a molecule of mass m has a mean kinetic energy proportional to

(A) $m^{1/2}$ (B) m (C) $m^{-1/2}$ (D) m^0

Knowledge Required: Effects on distribution properties as mass and temperature are changed.

Thinking it Through: The mean kinetic energy is given by the average value of $(½)ms^2$, or $(½)ms_{rms}^2$, where s_{rms} is the root mean squared speed. Because $s_{rms} = (3k_BT/m)^{1/2}$, $s_{rms}^2 = (3k_BT/m)$, and the average kinetic energy $= (½)m(3k_BT/m) = (^3/_2)k_BT$, which is independent of the mass of the molecule. Response **(D)** is thus the correct result.

KMT-11. The equipartition theorem tells us that, for *each* translational degree of freedom, there is a contribution to the average kinetic energy of

(A) $k_BT.$ (B) $(^3/_2) k_BT.$ (C) $(^1/_2) k_BT.$ (D) $(^1/_2) k_BT^2.$

Knowledge Required: Statistical treatment of energy levels and distribution of energy across types of energy.

Thinking it Through: In the last problem we showed that the average kinetic energy in 3 dimensions is $(3/2)k_BT$. Thus for each dimension, $(½)k_BT$ is contributed, and the correct response would be response **(C)**. Another way of remembering this fact is that each quadratic expression (translational energy, or the potential energy for a harmonic oscillator) contributes $(½)k_BT$ through the equipartition of energy.

Practice Questions

1. A plot describing the distribution of speeds in a gas is shown below, where $F(s) = \left(\dfrac{1}{N}\right)\left(\dfrac{dN}{ds}\right)$ is the Maxwell distribution function.

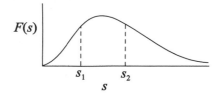

The fraction of molecules having speeds between s_1 and s_2 is given by

(A) $\int_{s_1}^{s_2} F(s)\,ds$ (B) $\int_{s_1}^{s_2} sF(s)\,ds$

(C) $\int_0^{s_2} F(s)\,ds$ (D) $\int_0^{\infty} sF(s)\,ds$

2. On the plot, the most probable speed is indicated by a, and the corresponding fraction of molecules is b. Increasing temperature causes

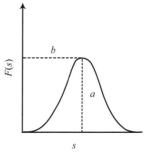

(A) both a and b to increase.

(B) both a and b to decrease.

(C) a to increase and b to decrease.

(D) a to decrease and b to increase.

3. Consider the speed distribution shown. As the temperature increases

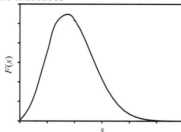

(A) the maximum stays at the same speed.

(B) the area under the curve increases.

(C) the distribution narrows.

(D) the maximum moves to the right.

4. In the gas phase, a molecule of mass m has mean speed proportional to

(A) $m^{1/2}$ (B) m

(C) $m^{-1/2}$ (D) m^0

5. Which is true for the distribution of molecular speeds (s)?

(A) $s_{\text{root mean square}} > s_{\text{mean}} > s_{\text{most probable}}$

(B) $s_{\text{most probable}} > s_{\text{root mean square}} > s_{\text{mean}}$

(C) $s_{\text{root mean square}} > s_{\text{most probable}} > s_{\text{mean}}$

(D) $s_{\text{root mean square}} = s_{\text{most probable}} > s_{\text{mean}}$

6. If the temperature is doubled, then the average speed of the molecules in an ideal gas will change by a factor of

(A) $(\tfrac{1}{2})^{1/2}$ (B) $\tfrac{1}{2}$

(C) $(2)^{1/2}$ (D) 2

7. The collision rate between molecules Kr and Ar in a 1 L bulb of gas filled with equal amounts of Kr and Ar

(A) increases with increasing pressure.

(B) decreases with increasing pressure.

(C) is independent of pressure.

(D) is the same as the collision frequency between Kr and Ar atoms.

8. In collision theory for the gas-phase reaction,

A + B → products

the reaction cross section, σ_r, is directly proportional to the rate constant. The value of this parameter is determined by the

(A) square of the diameter of A.

(B) square of the diameter of B.

(C) square of the sum of the radii for A and B.

(D) sum of the van der Waals radii of A and B.

9. The *most probable* energy of a molecule in a gas at temperature T is given by

(A) $(3/2)\,k_BT$.

(B) taking a statistical average of the energy distribution for the molecules.

(C) finding the maximum in the energy distribution function for the molecules.

(D) integrating the energy distribution function for the molecules.

10. Which set of relations is true concerning the plot of the speed distributions in one dimension for two gaseous samples 1 and 2 of mass M_1 and M_2?

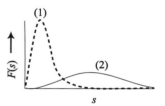

(A) $M_1 > M_2$ and $T_1 < T_2$

(B) $M_1 < M_2$ and $T_1 = T_2$

(C) $M_1 = M_2$ and $T_1 > T_2$

(D) $M_1 < M_2$ and $T_1 > T_2$

Answers to Study Questions

1. B
2. D
3. D
4. A
5. A
6. C
7. A
8. C
9. D
10. D
11. C

Answers to Practice Questions

1. A
2. C
3. D
4. C
5. A
6. C
7. A
8. C
9. C
10. A

Transport Properties

The transport properties are those properties involving transfer of molecules across boundaries. The idea of transport of materials in the gas phase, for example, has some familiar examples. On every trip you've ever taken to the movie theatre you've experienced them because you smell the popcorn the moment you walk in, though the snack counter is across the room. Somehow the molecules that your nose and brain interpret as the smell of popcorn were transferred from one area to another. In order to study this phenomenon in detail, we have to add the concept of a boundary, even an imaginary one (somewhere between you and the snack stand – at least we can hope it's an imaginary boundary!)

Recall from kinetic molecular theory, that gas molecules are in constant random motion and undergoing collisions with other molecules in the gas. Our models of transport properties need to incorporate these concepts and account for them in gases. Thus, we need to gain a feel for the distance a molecule travels between collisions as this distance is surely related to the transport properties. Because of the random nature of molecular motion, however, in most laboratory experiments the distance traveled between collisions is a variable and the most efficient way to describe the system is to consider the mean free path, or the average distance between collisions.

While gas phase systems provide a good starting point for this type of study, it is also important that we consider transport processes in liquids. Ultimately, chemical reactions are carried out more often in liquids and gases than in solids, so once we can understand transport in these phases we will have a good start in the study of kinetics. For liquid phases we consider the conductance and the Debye-Hückel theory for ionic solutions as examples of transport properties.
s
Finally, because it is possible to take advantage of transport properties to separate components of a chemical system we include a few problems that examine rates of processes involving the physical separation of materials.

Study Questions

TP-1.	Which property of a gas depends on the mean free path of the molecules?
	(A) the viscosity coefficient **(B)** the average molecular speed
	(C) the average molecular kinetic energy **(D)** the average momentum in the *x*-direction

Knowledge Required: Dependence of gas phase transport properties on molecular properties.

Thinking it Through: Responses **(B)**, **(C)**, and **(D)** are all properties of the distribution of molecular speeds or velocities and can not be a correct response. The viscosity coefficient does depend on the mean free path (λ) of the molecules as $\eta = (1/3) M \lambda \bar{s}$ [A]. Thus response **(A)** is correct.

TP-2.	For nitrogen gas at room temperature and pressure, the mean free path of the nitrogen molecules is approximately
	(A) 1 m **(B)** 10^{-3} m **(C)** 10^{-5} m **(D)** 10^{-10} m

Knowledge Required: Order of magnitude of gas phase collisional properties.

Thinking it Through: The mean free path is given by $\lambda = (\sigma(N/V))^{-1}$ within factors of the square root of 2. For N_2 at room temperature and pressure $(N/V) \approx 2.5 \times 10^{19}$ molecules cm^{-3} and $\sigma \approx 10^{-16}$ cm^2, thus $\lambda \approx [(10^{19})(10^{-16})]^{-1}$ = 10^{-3} cm or 10^{-5} m. Response **(C)** is correct.

Transport Properties

TP-3. The mean free path for a gas molecule depends on temperature T, molecular radius r, and pressure P. The mean free path is proportional to

(A) Tr^2P (B) $\frac{P}{Tr^2}$ (C) $\frac{T}{r^2P}$ (D) $\frac{r^2}{TP}$

Knowledge Required: Gas phase collisional properties and their dependence on fundamental molecular and macroscopic properties.

Thinking it Through: The mean free path is given by $\lambda = (\sigma(N/V))^{-1}$ within factors of the square root of 2. Substitution gives $\lambda = (\pi r^2(P/RT))^{-1}$ or in terms of r, P, and T; λ is proportional to $(r^2P/T)^{-1}$ or T/r^2P. Response **(C)** is correct.

TP-4. Order the gases nitrogen, oxygen, fluorine, and neon from slowest to fastest rate of effusion at 1 bar and 298 K.

(A) $F_2 < O_2 < N_2 < Ne$ (B) $N_2 < O_2 < F_2 < Ne$

(C) $Ne < N_2 < O_2 < F_2$ (D) $Ne < F_2 < O_2 < N_2$

Knowledge Required: Gas phase collisional properties and their dependence on fundamental molecular and macroscopic properties.

Thinking it Through: The rate of effusion is proportional to the inverse square root of the molar mass of the molecule. Thus the rate of effusion of the heaviest molecule is the slowest, and the lightest molecule effuses the fastest. Responses **(A)** and **(C)** order the molecules by molar mass. Response **(A)** from slowest to fastest (correct response) and response **(C)** from fastest to slowest (incorrect)

TP-5. At 25 °C the molar conductivity of Ag^+ ion is 61.90×10^{-4} m²·ohm⁻¹·mol⁻¹, while that of $AgNO_3$ is 133.36×10^{-4} m²·ohm⁻¹·mol⁻¹. The transport number for the Ag^+ ion is

(A) 71.46×10^{-4} (B) 0.4642

(C) 0.5358 (D) 2.1544

Knowledge Required: Definition of transfer number in terms of conductivities.

Thinking it Through: The transference number / transport number is the fraction of the current carried by a given ion. Thus the transference number for the silver ion is $\frac{\Lambda_{Ag^+}}{\Lambda_{Ag^+} + \Lambda_{NO_3^-}} = \frac{\Lambda_{Ag^+}}{\Lambda_{AgNO_3}} = \frac{61.90}{133.36} = 0.4642$ indicating that response **(B)** is the correct response.

TP-6. When the reaction
$$A^{2+} + 2X^- \rightarrow B^{2+} + Y^{2-}$$
is studied in pure water, aqueous 0.1 M NaCl, and 1.0 M NaCl, the apparent rate constants in these media would be in the order

(A) $k(\text{pure}) > k(1.0\text{ M}) > k(0.1\text{ M})$ (B) $k(0.1\text{ M}) > k(1.0\text{ M}) > k(\text{pure})$

(C) $k(1.0\text{ M}) > k(0.1\text{ M}) > k(\text{pure})$ (D) $k(\text{pure}) > k(0.1\text{ M}) > k(1.0\text{ M})$

Transport Properties

Knowledge Required: Effects of ionic species on rate constants, Debye – Hückel theory of ionic solutions.

Thinking it Through: The Debye-Hückel theory for ionic salts predicts that the rate constant for a reaction in ionic solution is given by

$$\log k = \log k^o + 2Az_Az_BI^{1/2}$$

This equation predicts that the rate constant will decrease as the square root of the ionic strength increases because of a negative slope from the opposite charges. The correct decreasing order is represented in response **(D)**.

TP-7. The molar conductivities (in units of ohm^{-1} cm^2 mol^{-1}) in infinite dilution (Λ_0) are given in the table. The value of Λ_0 for CH$_3$COOH(aq) in the same units is

Substance	Λ_0
HCl(aq)	426.1
NaCl(aq)	126.5
Na(CH$_3$COO)(aq)	91.0
H$^+$(aq)	349.8
OH$^-$(aq)	196.7

    (A) 214.5   **(B)** 390.6   **(C)** 517   **(D)** 643.6

Knowledge Required: Ionic solutions, molar conductivities of multi-component solutions.

Thinking it Through: The molar conductivity is an additive property of ionic solutions. From the given data, the molar conductivity at infinite dilution for acetic acid can be obtained from

$$\begin{aligned}\Lambda_0(\text{CH}_3\text{COOH}) &= \Lambda_0(\text{H}^+) + \Lambda_0(\text{CH}_3\text{COO}^-) \\ &= \Lambda_0(\text{H}^+) + (\Lambda_0(\text{NaCH}_3\text{COO}) - \Lambda_0(\text{Na}^+)) \\ &= \Lambda_0(\text{H}^+) + \Lambda_0(\text{NaCH}_3\text{COO}) - (\Lambda_0(\text{NaCl}) - \Lambda_0(\text{Cl}^-)) \\ &= \Lambda_0(\text{H}^+) + \Lambda_0(\text{NaCH}_3\text{COO}) - \Lambda_0(\text{NaCl}) + (\Lambda_0(\text{HCl}) - \Lambda_0(\text{H}^+)) \\ &= \Lambda_0(\text{H}^+) + \Lambda_0(\text{NaCH}_3\text{COO}) - \Lambda_0(\text{NaCl}) + \Lambda_0(\text{HCl}) - \Lambda_0(\text{H}^+) \\ &= 91.0 - 126.5 + 426.1 = 390.6\end{aligned}$$

Thus response **(B)** is the correct response.

TP-8. A particular *t*-RNA molecule takes 80 ms to diffuse 1 μm from the interior of a cell to the cell wall. How long will it take to diffuse 2 μm to the cell wall of a bigger cell?

    (A) 40 ms   **(B)** 80 ms   **(C)** 160 ms   **(D)** 320 ms

Knowledge Required: Diffusion rate dependence on distance, time, and molecular properties.

Thinking it Through: Diffusion is a process characterized by a Gaussian distribution of the distance of the molecules from their starting point. The distribution is characterized by the width of the distribution which is equal to $2(Dt/\pi)^{1/2}$, where D is the diffusion coefficient and t is the time. Thus the distance diffused is proportional to t^2. If a molecule diffuses 1 μm in 80 ms then it will take 4 times as long (or 320 ms) to diffuse 2 μm, thus response **(D)** is the correct response.

TP-9. The rate of sedimentation of a molecule in an ultracentrifuge does not depend on

    (A) viscosity of the solvent.   **(B)** speed of the centrifuge.

    (C) shape of the molecule.   **(D)** time.

Transport Properties

Knowledge Required: Sedimentation rate dependence on parameters.

Thinking it Through: The sedimentation rate will depend on the viscosity of the solvent by slowing down as the viscosity increases, thus response **(A)** is incorrect. The speed of the centrifuge influences the force on the molecules, which changes the rate of sedimentation, thus response **(B)** is incorrect. If you consider a cylindrical molecule, the rate that the molecule moves through the solvent will depend on the area of the cylinder (if tilted, this increases) moving down through the solvent, thus response **(C)** is incorrect. This leaves response **(D)** to be the correct answer.

TP-10.	If E is the energy required to break up a cluster of molecules in a liquid, then the viscosity, η, of the liquid is approximately

(A) proportional to $e^{(E/RT)}$. (B) proportional to $e^{(-E/RT)}$.

(C) proportional to T^2. (D) independent of temperature.

Knowledge Required: Scaling of energies for processes, energy distributions.

Thinking it Through: The probability that a molecule has energy greater than E is approximately proportional to $e^{-E/RT}$, which eliminates the possibility of responses **(C)** or **(D)** being correct. E/RT will decrease as temperature increases, thus $e^{E/RT}$ will decrease as temperature increases and $e^{-E/RT}$ will increase as temperature increases. Experimentally, the viscosity decreases as temperature increases, making the viscosity proportional to $e^{E/RT}$ and response **(A)** the correct answer.

Practice Questions

1. The transport number for H⁺ in an aqueous solution of HCl can be estimated from the data given as

Substance	Λ_0
HCl(aq)	426.1
NaCl(aq)	126.5
Na(CH₃COO)(aq)	91.0
H⁺(aq)	349.8
OH⁻(aq)	196.7

(A) 0.3599 (B) 0.5623
(C) 0.8209 (D) 1.7783

2. The mobilities of H⁺ and Cl⁻ in water are 3.62×10^{-3} cm² · s⁻¹ · V⁻¹ and 7.91×10^{-4} cm² · s⁻¹ · V⁻¹ respectively. When a voltage is applied across an HCl solution, the percentage of the current carried by the positive charge is

(A) 82.1 % (B) 50%
(C) 31.3% (D) 17.9%

3. Which plot best represents the concentration dependence of the molar conductance for dilute strong electrolytes?

(A) 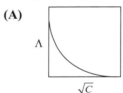 (B)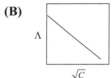

(C) 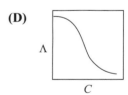 (D)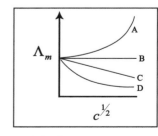

4. Which plot best represents the concentration dependence of the molar conductivity of a weak electrolyte?

(A) A (B) B
(C) C (D) D

5. The mean free path for a spherical gas molecule is inversely proportional to

   (A) the radius of the molecule.
   (B) the diameter of the molecule.
   (C) the square root of the radius of the molecule.
   (D) the square of the diameter of the molecule.

6. The units for heat flux are

   (A) $J \cdot s^{-1} \cdot m^2$
   (B) s^{-1}
   (C) $J \cdot s^{-1}$
   (D) $J \cdot s^{-1} \cdot m^{-2}$

7. In general, for ions reacting in solution, the rate

   (A) increases between ions of like charge as ionic strength increases.
   (B) decreases between ions of like charge as ionic strength increases.
   (C) increases between ions of opposite charge as ionic strength increases.
   (D) is independent of ionic strength.

8. The mean free path of the molecules in a gas at 1 atm pressure and 25°C is

   (A) directly proportional to the square of their diameter.
   (B) the same as at 10 atm pressure.
   (C) inversely proportional to their collision frequency.
   (D) independent of their diameter.

9. In the design (or construction) of a vacuum system, large diameter tubing is essential for the main manifold. The property of gas molecules being considered in this instance is

   (A) collision diameter.
   (B) root mean square velocity.
   (C) mean free path.
   (D) average molecular mass.

10. Which of the properties of a gas depends on the mean-free path of the molecules?

    (A) the viscosity coefficient
    (B) the average molecular speed
    (C) the average molecular kinetic energy
    (D) the average momentum in the x-direction

Answers to Study Questions

1. A
2. C
3. C
4. A
5. B
6. D
7. B
8. D
9. D
10. A

Answers to Practice Questions

1. C
2. A
3. B
4. D
5. D
6. D
7. A
8. C
9. C
10. A

Phenomenological Kinetics

While we started our consideration of kinetics by thinking about how molecular level interactions occur, it's important to realize that the study of reaction rates actually began with experimental studies at the macroscopic level. Once a large number of experiments were conducted patterns began to emerge about ways that they might be described, particularly in terms of mathematical models. Because these studies and the mathematics describing them focus on the observable phenomena of reaction kinetics, they are categorized as phenomenological kinetics.

There are a number of factors that influence the rate of reactions, and among the first to be studied systematically was the effect of the concentration (or amount) of substances present. Studies of the concentration dependence of kinetics resulted in the definition of both the rate and the rate law.

The fundamental premise of a rate law is that the rate is proportional to the concentration of reactants raised to some small, commonly integer, power.

$$\text{Rate} \propto [X]^y$$

where we are designating the reactant of interest as X and the exponent y is the small integer (or half integer.) These integers can only be determined by experiments, but once they are known, the rate law provides a wealth of information about the time behavior of the reaction it describes. Given its utility it is not surprising that there are actually multiple ways to determine the rate law and this chapter provides several questions to provide some breadth of experience in this form of kinetics problem.

The determination of a rate law is important mostly because once we know it we can make predictions about the progress of the reaction. For an individual reaction, we would most likely wish to know an expected concentration at a given time, but the fact that the rate law is similar for many reactions allows us to use this idea to compare rates of reactions. There are several ways that this might be accomplished, but perhaps the most common is to determine the half-life of the reaction. The half-life is defined to be the amount of time it takes for the concentration (or amount) of a limiting reactant to fall to one-half its initial value.

Concentration of reactants is not the only variable that can be manipulated in a kinetics experiment, however. Common experience also tells us that temperature plays a role in how fast a reaction takes place. At higher temperatures reactions usually proceed more quickly. At the macroscopic level, the temperature dependence of kinetics is generally described by an equation attributed to Svante Arrhenius,

$$k = A e^{-E_a/RT}$$

Our final questions look at Arrhenius behavior of chemical kinetics.

Study Questions

PK-1. For the reaction $2NO_2(g) + F_2(g) \rightarrow 2NO_2F(g)$ the rates of changes of concentrations are related by:

(A) $(½) \, d[NO_2]/dt = d[F_2]/dt$

(B) $(2) \, d[NO_2]/dt = d[F_2]/dt$

(C) $d[NO_2]/dt = d[NO_2F]/dt$

(D) $d[F_2]/dt = -d[NO_2F]/dt$

Phenomenological Kinetics

Knowledge Required: Rate laws, definition of reaction rate

Thinking it Through: The definition of the rate is given by

$$\text{Rate} = \frac{1}{\upsilon_i}\frac{dc_i}{dt}$$

where the υ_i are the stoichiometric coefficients for species i (positive for products and negative for reactants) and c_i is the concentration of species i. Thus, for this reaction

$$\text{Rate} = -\frac{1}{2}\frac{d[NO_2]}{dt} = -\frac{1}{1}\frac{d[F_2]}{dt} = \frac{1}{2}\frac{d[NO_2F]}{dt}$$

so that response **(A)** is correct. Response **(B)** is incorrect because the 2 is not in the denominator. Response **(C)** is incorrect because there should be a negative sign present, and response **(D)** is incorrect because of the missing factor of 2.

PK-2. Initial rates, R_0, for a reaction A + B $\rightarrow$ products, which takes place in the presence of a catalyst C, were measured at 298 K for different initial concentrations of A, B and C (in mol dm^{-3}). Assuming that the rate equation has the form

$$\text{Rate} = k\,[A]^\alpha\,[B]^\beta\,[C]^\gamma$$

$[A]_0$	$[B]_0$	$[C]_0$	R_0
0.40	0.30	0.60	0.60
1.20	0.30	0.60	1.79
0.40	0.30	1.80	5.41
1.20	0.90	0.60	1.79

The values of α, β and γ are:

(A) α = 1   β = 0   γ = 3
(B) α = 1   β = 0   γ = 2
(C) α = 2   β = 1   γ = 2
(D) α = 3   β = 1   γ = 2

Knowledge Required: Determination of rate laws from multiple experiments with concentration-rate data.

Thinking it Through: The form of the rate law is given in the problem. Identifying the four experiments as 1, 2, 3, and 4 from the top, we can write

$$\frac{\text{Rate}_i}{\text{Rate}_j} = \frac{k[A]^\alpha_{0,i}[B]^\beta_{0,i}[C]^\gamma_{0,i}}{k[A]^\alpha_{0,j}[B]^\beta_{0,j}[C]^\gamma_{0,j}}$$

where the i, j's correspond to the different experiments. Selecting experiments 1 and 2 allows cancellation of the terms involving k, [B], and [C] to give

$$\frac{1.79}{0.60} = \left(\frac{1.20}{0.40}\right)^\alpha$$

and α = 1. Thus responses **(C)** and **(D)** are incorrect. β is equal to 0 is both responses **(A)** and **(B)**, thus γ must be used to differentiate between the two responses. Selecting experiments 1 and 3 allows cancellation of the terms involving k, [A], and [B] to give

$$\frac{5.41}{0.60} = \left(\frac{1.80}{0.60}\right)^\gamma$$

from which γ is found to be 2, thus response **(B)** is correct and response **(A)** is incorrect.

PK-3. Which plot of concentration-time data will produce a straight line for a zeroth order reaction?

(A) c^{-1} against t (B) c against t (C) c^{-1} against t^{-1} (D) $\ln c$ against t

Knowledge Required: Linearization of concentration-time data for various orders.

Thinking it Through: For a zeroth order reaction, the time dependence (integrated rate law) is given by

$$c = c_0 - kt$$

Thus a plot of c (y axis) against t (x axis) will be linear with intercept c_0 and slope $-k$. This leads to response **(B)** being correct. Response **(A)** would be correct for a second order reaction, response **(C)** is always incorrect (t^{-1} on x axis is not used), and response **(D)** would be correct for a first order reaction.

PK-4. For the reaction, A $\rightarrow$ products, a plot of the concentration of A vs. time is linear. What is the order of the reaction?

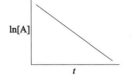

(A) zero (B) first (C) second (D) third

Knowledge Required: Linearization of concentration-time data for various orders.

Thinking it Through: The integrated rate law expressions for zeroth through third order kinetics of

$[A]-[A]_0 = -kt$, $\ln\dfrac{[A]}{[A]_0} = -kt$, $\dfrac{1}{[A]} - \dfrac{1}{[A]_0} = kt$,   and $\dfrac{1}{[A]^2} - \dfrac{1}{[A]_0^2} = 3kt$, respectively. Thus the plot shows $\ln[A]$ against t as linear; thus the reaction is first order and the correct response is **(B)**.

PK-5. Consider a reaction that is first order in both reactants A + B $\rightarrow$ products. The reaction becomes a pseudo-first order reaction if

(A) the temperature is raised by 10 K. (B) a catalyst is used.
(C) reactant B is present is large excess. (D) equilibrium is established.

Knowledge Required: Rate laws, definitions of pseudo-order reaction conditions, how to make reaction become pseudo-order.

Thinking it Through: The rate law for a reaction that is first order in both reactants is given by

$$\text{Rate} = k\,[A]^1\,[B]^1$$

In order to become a pseudo-first order reaction, either A or B must be present in large excess so that its concentration becomes invariant. Response **(C)** gives the case where B is present in large excess, and is thus the correct response. In response **(A)**, the small temperature changes typically do not affect the rate law and is thus incorrect. In response **(B)**, a catalyst typically affects the rate law by increasing the complexity of the rate law not by making it pseudo-order, thus is incorrect. The equilibrium response of response **(D)** has nothing to do with the rate law, and is thus incorrect.

Phenomenological Kinetics

PK-6. Femtosecond spectroscopy can be used to examine reactions on the 10^{-15} second time scale. These experiments are most useful for

(A) reactions that are faster than a single vibration.

(B) diffusion controlled reactions.

(C) typical laboratory organic reactions.

(D) reactions that require adsorption on a surface.

Knowledge Required: Experimental techniques for types of reactions.

Thinking it Through: One femtosecond is 10^{-15} s. Response **(A)** compares this time with that of molecular vibration (10^{-12} seconds). Response **(B)** compares this time with that of diffusion controlled reaction ($10^{-3} - 10^{-9}$ seconds). Response **(C)** compares this time with that of your organic laboratory reactions (minutes to hours). Response **(D)** compares this time with that of reactions requiring adsorption on a surface (requires colliding with the surface which is of order 10^{-6} seconds). The shortest of these time scales is that of response **(A)**, which is the correct response.

PK-7. The relationship between rate constant k, initial concentration c_0, and half-life $t_{1/2}$, for a zeroth-order reaction is

(A) $t_{1/2} = 1/kc_0$ (B) $t_{1/2} = (\ln 2)/k$ (C) $t_{1/2} = kc_0$ (D) $t_{1/2} = c_0/2k$

Knowledge Required: Definition of half-life, dependence of half-life on order and concentration.

Thinking it Through: The general definition of the half-life is the time required for the concentration to decrease to ½ of the original amount. For a zeroth order reaction, the time dependence (integrated rate law) is given by

$$c = c_0 - kt$$

Substitution of $c(t_{1/2})$ for $c(t)$ and $t = t_{1/2}$ gives

$$c_0/2 = c_0 - k\, t_{1/2}$$

and $t_{1/2} = c_0/2k$ (response **(D)** is correct).

PK-8. Two first-order reactions have identical pre-exponential factors; their activation energies differ by 25.0 kJ·mol^{-1}. The ratio of their rate constants at 25°C is

(A) 50 (B) 2.4×10^4 (C) 6.1×10^4 (D) 1.7×10^{52}

Knowledge Required: Arrhenius explanation of temperature dependence of reactions.

Thinking it Through: Since we are examining the effects of the activation energies on two reactions at one temperature, we need to examine the ratio of the rate constants. This ratio is

$$\frac{k_1}{k_2} = \frac{A_1 e^{-E_{a,1}/RT}}{A_2 e^{-E_{a,2}/RT}}$$

Since $A_1 = A_2$, this ratio can be rewritten as

$$\frac{k_1}{k_2} = e^{-(E_{a,1} - E_{a,2})/RT} = e^{\pm 25{,}000/((8.314)(298))} = 4.1 \times 10^{-5} \text{ or } 2.4 \times 10^4$$

Looking at the two values, 4.1×10^{-5} is so small that it must be the answer for $\dfrac{k_2}{k_1}$ ratio and it is not available as an answer choice, so **(B)** is the correct response.

Phenomenological Kinetics

PK-9. For the concerted reaction $A + B \rightarrow C + D$, E_a(forward) is 46 kJ·mol^{-1} and $\Delta H = -27$ kJ·mol^{-1}. Therefore, E_a (in units of kJ·mol^{-1}) for the reverse reaction is

(A) 19   (B) 27   (C) 46   (D) 73

Knowledge Required: Diagram of energy as a function of reaction coordinate, relations between energies of reactants, products and the transition state

Thinking it Through: The conventional diagram of the energy (y) against reaction coordinate (x) shows the energy difference between reactants and products as ΔH. There is a barrier to the reaction between reactants and products. The height of this barrier corresponds to the activation energy for the forward reaction when moving to the right, and the activation energy for the reverse reaction when moving to the left. In this case, the height of the barrier for the reverse reaction E_a(reverse) = 46 kJ·mol^{-1} + 27 kJ·mol^{-1} = 73 kJ·mol^{-1}. The correct answer is response **(D)**. Response **(A)** corresponds to (46 – 27) kJ·mol^{-1}, which is incorrect. Responses **(B)** and **(C)** correspond to the two numbers given in the problem, also incorrect responses.

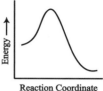

PK-10. The gas-phase reaction $2NO_2 + O_3 \rightarrow N_2O_5 + O_2$ has the rate constant $k = 2.0 \times 10^4$ L·mol^{-1}·s^{-1} at 300 K. What is the order of the reaction?

(A) 0   (B) 1
(C) 2   (D) Unable to determine with the data given.

Knowledge Required: Units as indicators of reaction orders

Thinking it Through: The rate law expression is given by $\dfrac{d[A]}{dt} = -k[A]^x$. Thus the units on the rate constant, k, are mol·L^{-1}·time^{-1}/(mol·L^{-1})^x. Since the given units of the rate constant are L·mol^{-1}·s^{-1} the value of x (the order) is 2, and response **(C)** is correct.

Practice Questions

1. When the time dependence of the reaction

    $A + B \rightarrow C$

    was studied at different initial [A] and [B], the relative rates were shown in the table.

[A]	[B]	Rate
c_0	c_0	R_0
$2c_0$	c_0	$4R_0$
$2c_0$	$2c_0$	$8R_0$

    From this data one concludes that the overall rate for this reaction is

    (A) zero order.   (B) first order.
    (C) second order.   (D) third order.

2. The chemical reaction

    $A + B \rightarrow C$

    is first order in A and first order in B. When the initial concentrations of A and B are both equal to c_0, the initial rate for the reaction is proportional to

    (A) a constant.   (B) c_0.
    (C) c_0^3.   (D) c_0^2.

Phenomenological Kinetics

3. The chemical reaction, A → products, is found to have a rate equation $\frac{d[A]}{dt} = -k[A]^2$. The half life for this reaction is

   (A) independent of [A].
   (B) proportional to 1 / [A].
   (C) equal to k.
   (D) zero.

4. For the reaction A → B the concentration of A varies as shown. The reaction must be first order with respect to [A] because

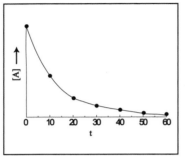

   (A) the half-life of the reaction is constant.
   (B) the half-life of the reaction is inversely proportional to [A].
   (C) the half-life of the reaction is proportional to [A].
   (D) [A] is independent of the half-life.

5. For $2H_2O_2(aq) \rightarrow 2H_2O(l) + O_2(g)$, which plot confirms that the reaction is second order with respect to H_2O_2?

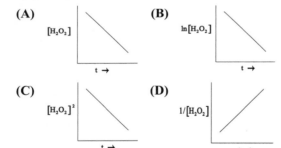

6. If the rate constant for a chemical reaction is k_0 at 300 K, the activation energy is E_a (kJ·mol^{-1}), and the pre-exponential factor, A, is not a function of temperature, then the rate constant at 310 K is

   (A) $k_0\exp[-(E_a/R)(1/310 - 1/300)]$
   (B) $k_0\exp[-(E_a/R)(1/310 + 1/300)]$
   (C) $k_0\exp[-(E_a/R)(1/300 - 1/310)]$
   (D) $k_0\exp[(E_a/R)(1/310 + 1/300)]$

7. Under what conditions will the temperature dependence of the rate constant (dk/dT) be greatest?

   (A) low temperature, small E_a
   (B) low temperature, large E_a
   (C) high temperature, small E_a
   (D) high temperature, large E_a

8. Which correctly expresses the ratio of the rate constant at 30 °C to that at 25 °C?

   $k_{30} / k_{25} =$

   (A) $303^{-1} - 298^{-1}$
   (B) $30^{-1} - 25^{-1}$
   (C) $\exp[-(E_a/R)/(303^{-1} - 298^{-1})]$
   (D) $\exp[-(E_a/R)/(30^{-1} - 25^{-1})]$

9. For a second order chemical reaction, the units for the rate constant (k) are

   (A) (concentration)(time)$^{-1}$
   (B) (concentration)^2(time)$^{-1}$
   (C) (concentration)$^{-2}$(time)$^{-1}$
   (D) (concentration)$^{-1}$(time)$^{-1}$

10. A first order reaction in solution might be followed by absorbance measurements. Which quantities would have to be measured to determine the rate constant, assuming the absorption spectra of the reactants and products do not overlap?

    (A) The absorbance at the start of the reaction
    (B) The concentration of the starting material
    (C) The extinction coefficient or molar absorptivity of each reactant and product
    (D) The absorbance of the reactant at two different times

Answers to Study Questions

1. A
2. B
3. B
4. B
5. C
6. A
7. D
8. B
9. D
10. C

Answers to Practice Questions

1. D
2. D
3. B
4. A
5. D
6. A
7. B
8. C
9. D
10. D

Mechanisms

If we think about the nature of molecular collisions as devised by kinetic molecular theory we realize that reactions frequently proceed in a series of steps. The probability of a type of collision occurring depends on the number of molecules involved. The most common is a two-molecule collision. Three molecules can collide occasionally but that would probably be rare and four molecule collisions must be so rare as to be unimportant. Nonetheless, stoichiometric depictions of reactions would seem to imply the involvement of many molecules at a time in at least some reactions. This mismatch between collisions and stoichiometry implies that reactions must be able to proceed in steps, something we refer to as the mechanism of the reaction.

Mechanisms are a series of elementary reactions proposed to explain how a reaction occurs. Elementary reactions are labeled by the number of molecules that collide in them.
- One molecule involved – unimolecular
- Two molecules involved – bimolecular
- Three molecules involved – termolecular

Mechanisms can be disproved by experimental evidence, but it is not possible to prove them to be true because it is always possible that another mechanism could be devised that accounts for the same observations. There are, however, important constraints that must be met for any mechanism to be accepted as reasonable. Foremost, the steps of the mechanism must sum to yield the net reaction being observed. As might be expected from our initial discussion here, the mechanism cannot have any reaction that would imply more than a termolecular step and practically speaking most steps are unimolecular or bimolecular.

In addition to being able to describe the overall reaction stoichiometry, the mechanism must also be able to explain the observed rate law. There are several factors that affect the manner in which a mechanism can be distilled into a rate law. For example, some problems will review approximation methods such as the steady state approximation and the fast equilibrium approximation. Other problems reviewing rate limiting steps and comparison of rate constants for elementary steps to simplify rate laws are also included.

Another area where reaction mechanisms provide an important way to extend our understanding of kinetics lies in the concept of catalysis. A catalyst enhances the rate of a reaction without actually being consumed as the reaction proceeds, and the mechanism needs to account for this effect. Catalysis has many practical implications and we will review reaction mechanisms important in practical cases such as enzyme kinetics, photochemical reactions and chain reactions (such as polymerization).

Study Questions

M-1. Consider the reaction mechanism $\boxed{A \rightarrow B \rightarrow C}$ in which the rate constant for the first reaction is very much smaller than that for the second reaction. At $t = 0$, only substance A is present. Shortly after the steady-state is initially established (both reactions are irreversible), which of the following statements is correct?

(A)   [C] < [B]   (B)   [B] ≈ 0   (C)   [A] < [C]   (D)   [B] ≈ [A]

Knowledge Required: Consecutive reactions, interpreting reactions using rate-limiting steps

Thinking it Through: Because the rate constant for the first reaction is small (reaction is slow), A will slowly convert to B. The rate constant for the second reaction is large (reaction is fast), therefore as soon as any B forms it is quickly converted to C. This means that the concentration of B will be zero or very small after the reaction reaches a steady-state, and response **(B)** is correct.

M-2. The sequence

$2 CH_3NC \rightarrow CH_3NC^* + CH_3NC$		k_1
$CH_3NC^* + CH_3NC \rightarrow 2 CH_3NC$		k_{-1}
$CH_3NC^* \rightarrow CH_3CN$		k_2

is proposed to account for the first-order kinetics observed in the $CH_3NC(g) \rightarrow CH_3CN(g)$ isomerization at high pressures. Which assumptions are made to account for the observed first-order rate law,

$d[CH_3NC] / dt = - k[CH_3NC]$

- I Apply the steady state approximation to CH_3NC^*
- II Assume $k_{-1} [CH_3NC] \gg k_2$
- III Assume $k_1 \gg k_2$
- IV Assume $k_1 / k_{-1} \gg 1$

(A) I only **(B)** I and II **(C)** I, II, and III **(D)** I, II, III, and IV

Knowledge Required: Writing rate expressions using the elementary steps in a mechanism. Identifying intermediates and applying the steady-state approximation. Simplifying rate expressions by comparing the magnitude of individual rate constants.

Thinking it Through: CH_3NC^* is an intermediate. Write a rate expression for the rate of change in concentration of this intermediate and apply the steady-state approximation:

$$\frac{d[CH_3NC^*]}{dt} = 0 = k_1[CH_3NC]^2 - k_{-1}[CH_3NC^*][CH_3NC] - k_2[CH_3NC^*]$$

Solve this expression for the concentration of the intermediate:

$$[CH_3NC^*] = \frac{k_1[CH_3NC]^2}{k_{-1}[CH_3NC] + k_2}.$$

Now write a rate expression for the change in concentration of CH_3NC:

$d[CH_3NC]/dt = k_2[CH_3NC^*]$.

Substitute for the concentration of the intermediate found by applying the steady-state approximation

$$\frac{d[CH_3NC]}{dt} = \frac{k_1 k_2 [CH_3NC]^2}{k_{-1}[CH_3NC] + k_2}$$

If this expression is to be reduced to a first order expression, k_2 must be small in comparison to the other term in the denominator. Therefore the correct response is **(B)** (conditions I and II are applied).

M-3. The rate expression for the decomposition of a substrate (S) in the presence of an enzyme (E) is

$$\text{rate} = \frac{k_2 [E]_0 [S]}{K_M + [S]}$$

As [S] becomes very large compared to K_M, the apparent order for the substrate is

(A) zero **(B)** one **(C)** two **(D)** two-thirds

Knowledge Required: Simplifying rate expressions. Determining partial reaction orders from rate laws.

Thinking it Through: If $K_M \ll [S]$, the denominator of the rate expression is approximately equal to [S]. The rate law becomes rate = $k [E]_0$. This rate expression does not include [S] and is therefore zero order with respect to the substrate, and response **(A)** is correct.

Mechanisms

M-4. In the mechanism shown, Step 1 is irreversible. Step 2 is reversible. At equilibrium

$$A \xrightarrow{k_1} B \underset{k_{-1}}{\overset{k_1}{\rightleftharpoons}} C$$

(A) [A] = 0 (B) [B] = 0 (C) [C] = 0 (D) [B] = [C]

Knowledge Required: Consecutive reactions, reversible reactions and equilibrium

Thinking it Through: Since the first reaction is not reversible, there will not be any A left at equilibrium. The second reaction is reversible, so some C and B will be present at equilibrium. Their relative amounts will depend on the value of the equilibrium constant; they will not necessarily be equal in concentration. Therefore response **(A)** is the best answer.

M-5. In homogeneous catalysis of a chemical reaction, the introduction of the catalyst changes the

(A) ΔH for the reaction.
(B) pathway between the products and reactants.
(C) equilibrium position of the reaction.
(D) number of products.

Knowledge Required: Definition of a catalyst

Thinking it Through: Catalysts affect the kinetics of a reaction, by providing a lower energy pathway for a reaction. Catalysts are not produced or consumed in a reaction and do not affect the stoichiometry of a reaction. Therefore response **(B)** is the correct answer.

M-6. When dissociated single strands, SS, of DNA's double helix recombine, the behavior shown in the figure is observed. It can be inferred that the reaction is initially _____ order, and that at time t_x _____

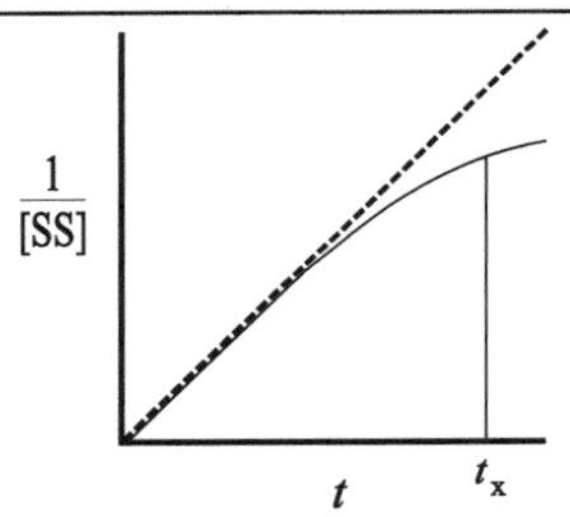

(A) first   equilibrium is established.
(B) first   the mechanism has changed.
(C) second   equilibrium is established.
(D) second   the mechanism has changed.

Knowledge Required: Using plots of integrated rate laws to determine reaction order.

Thinking it Through: Linear plots of 1/concentration against time indicate that a reaction is second order. Therefore this reaction is initially second order. At longer times (longer than t_x) the plot is no longer linear indicating a change in mechanism. Therefore response **(D)** is the best answer.

M-7. A molecule has absorbed a photon. In general, which process has the slowest rate for decay?

(A) stimulated emission
(B) phosphorescence
(C) internal conversion
(D) fluorescence

Knowledge Required: Understanding the differences between different types of electronic emission.

Thinking it Through: Response **(B)** is the best answer. Because an intersystem crossing occurs in phosphorescence this process has the slowest decay rate. The energy is trapped temporarily in the excited state of a forbidden transition.

M-8. Which sequence of state-to-state transitions for a diatomic system is required for one to observe fluorescence with a wavelength longer than that for the excitation photon?

(A)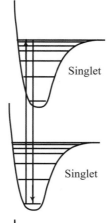

(B)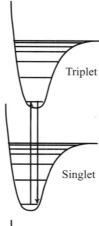

(C)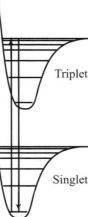

(D)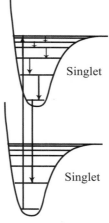

Knowledge Required: Fluorescence, energy diagrams, relationship between wavelength and energy

Thinking it Through: State-to-state fluorescence transitions occur between electronic states with the same multiplicity. Therefore responses **(B)** and **(C)** are not possible. Figure **(A)** represents a transition where the wavelengths are equal for the fluorescence and excitation photons, because the transition begins and ends in the same state. Figure **(D)** is the correct response. The fluorescence wavelength is longer (less energy) than that of the excitation photon.

M-9. A phosphorescent crystal was illuminated with a high intensity light beam. At $t = 0$, the light beam was turned off and the intensity of the phosphorescent radiation was measured in intervals of milliseconds. The equation that best describes the phosphorescent intensity(I) as a function of time(t) is

(A) $I = I_0$

(B) $\ln(I) = \ln(I_0) - kt$

(C) $I = I_0 - kt$

(D) $I = \ln(I_0) + kt$

Knowledge Required: Phosphorescence

Thinking it Through: Phosphorescence is first order with respect to intensity. Response **(B)** is the only answer that corresponds to a first order process, and is therefore the correct response.

Mechanisms

M-10. The photochemical reduction-dimerization of benzophenone to benzpinacol in isopropyl alcohol proceeds according to the equation with a quantum yield of 2.

$$2\ \phi_2C=O + (CH_3)_2CHOH \xrightarrow[300-350\ nm]{h\nu} HO-C(\phi)_2-C(\phi)_2-OH + (CH_3)_2C=O$$

The addition of a small amount of naphthalene to this reaction results in a drastic reduction in quantum yield although naphthalene does not absorb light in the region 300 – 350 nm. How do you explain this effect?

(A) Naphthalene reacts with benzophenone in its ground state and prevents its photochemical excitation.

(B) Naphthalene quenches reaction by transferring energy from excited benzophenone causing its return to the ground state.

(C) Excited naphthalene radicals terminate reaction by reacting with isopropyl alcohol radicals.

(D) Naphthalene is excited by radiation in preference to benzophenone.

Knowledge Required: Quantum yield, photochemical reactions

Thinking it Through: The reaction in question proceeds through the production of an excited state of benzophenone. The addition of naphthalene with its large number of π electrons and relative high mass will quench the excited state to the ground state. Thus response **(B)** is correct.

Practice Questions

1. For the reaction of A with B to form D, the first step involves forming a reaction intermediate C which is in equilibrium with A and B. The overall mechanism can be written as shown.

$$A + B \underset{k_{-1}}{\overset{k_1}{\rightleftharpoons}} C$$

$$C + B \xrightarrow{k_2} D$$

The rate expression for the production of D from this mechanism is expected to be

(A) (constant) [A].

(B) (constant) [A] [B].

(C) [A] [B] [C]^2.

(D) (constant) [A] [B]^2.

2. The hydrolysis of a certain transition metal complex can be written schematically as shown.

A → B

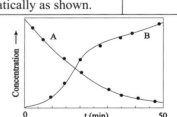

Which can be inferred from the figure?

(A) $-\left(\dfrac{d[A]}{dt}\right) = \left(\dfrac{d[B]}{dt}\right)$ at $t = 0$.

(B) At least one intermediate is formed.

(C) The rate of formation of B is a maximum at $t = 0$.

(D) The rate reaches equilibrium in 50 minutes.

Mechanisms

3. The predicted rate law for

$$A + B \underset{k_{-1}}{\overset{k_1}{\rightleftharpoons}} C \qquad C \xrightarrow{k_2} D$$

using the steady-state approximation is

(A) $\dfrac{k_1 k_2 [A][B]}{k_{-1}+k_2}$

(B) $k_2[C]$

(C) $k_1[A][B] - k_{-1}[C] - k_2[C]$

(D) $(k_1 k_2 / k_{-1})[A][B]$

4. Using the mechanism for free radical polymerization shown,

I → 2I·	k_1
I· + M → P·	k_2
P_n· + M → P_{n+1}	k_3
P_n· + P_m· → P_{m+n}	k_4

the kinetics are most easily modeled by applying the steady state approximation to

(A) initiator (I).

(B) monomer (M).

(C) polymer of chain length n (P_n).

(D) total polymer radical chain concentration ($P_n + P_m + \ldots$).

5. A possible mechanism for the gas-phase reaction

$$2NO_2Cl \rightarrow 2NO_2 + Cl_2$$

is:
| $NO_2Cl \rightarrow NO_2 + Cl$ (slow) |
| $Cl + NO_2Cl \rightarrow NO_2 + Cl_2$ (fast) |

The rate equation consistent with this mechanism is:

(A) rate = k [NO$_2$Cl]

(B) rate = k [NO$_2$Cl]^2

(C) rate = k [Cl] [NO$_2$Cl]

(D) rate = k [Cl$_2$] [NO$_2$]^2

6. The mechanism of an enzyme catalyzed reaction of a substrate (S) to yield products (P) can be written

$$E + S \rightleftharpoons ES \rightleftharpoons EP \rightleftharpoons P + E$$

where ES and EP represent the enzyme-substrate and enzyme-product complexes, respectively. For this mechanism, the potential energy diagram (potential energy or enthalpy against reaction coordinate) would be expected to have how many maxima?

(A) one   (B) two

(C) three   (D) four

7. Which statement is true when the temperature at a reaction mixture is increased?

(A) Temperature increases cause the equilibrium constant and final product concentrations to increase.

(B) A temperature increase reduces the activation energy for a reaction.

(C) A temperature increase speeds up the rate of a reaction by increasing the number of collisions.

(D) A temperature increase increases the rates of both the forward and reverse reactions.

8. According to the Langmuir model of chemisorption of a gas on a solid surface,

(A) the amount of gas adsorbed increases linearly with the partial pressure of the gas up to the saturation pressure.

(B) the amount of gas adsorbed per unit area of the surface approaches a limit as the partial pressure of the gas increases.

(C) a gaseous molecule striking a empty site on the surface has the same probability of being adsorbed as if it struck a site occupied by another molecule.

(D) the enthalpy change on adsorption is about the same as the enthalpy change on liquefaction of the gas.

Mechanisms

9. For the photochemically driven reaction of H_2 with I_2, a possible mechanism is shown in the box.

$$I_2 + h\nu \rightarrow 2I\cdot$$
$$I\cdot + H_2 \rightarrow HI + H$$
$$H + I_2 \rightarrow I\cdot + HI$$
$$I\cdot + I\cdot \rightarrow I_2$$

With this mechanism the number of HI molecules produced per photon absorbed is

(A) greater than 1.
(B) less than 1.
(C) equal to 1.
(D) always a whole number.

10. Gaseous, unimolecular decomposition reactions may proceed according to the mechanism in which the collision between two normal molecules, A, produces an activated molecule, A*, which, in turn, may be deactivated by collision or decompose into products.

$$A + A \xrightarrow{k_1} A^* + A$$
$$A^* + A \xrightarrow{k_{-1}} A + A$$
$$A^* \xrightarrow{k_2} \text{products}$$

Using the steady state approximation for the quantity $d[A^*]/dt$, the rate law for the reaction A $\rightarrow$ products becomes

$$-d[A]/dt = \frac{k_2 k_1 [A]^2}{k_{-1}[A] + k_2}$$

Under which conditions will the rate of the overall reaction tend to be second order?

(A) High pressure
(B) Low pressure
(C) Addition of inert gas
(D) Increased surface area in reaction vessel

11. For the mechanism

$$CH_3COCH_3 \xrightarrow{k_1} CH_3\cdot + CH_3CO\cdot$$
$$CH_3CO\cdot \xrightarrow{k_2} CH_3\cdot + CO$$
$$CH_3\cdot + CH_3COCH_3 \xrightarrow{k_3} CH_4 + \cdot CH_2COCH_3$$
$$\cdot CH_2COCH_3 \xrightarrow{k_4} CH_3\cdot + CH_2CO$$
$$CH_3\cdot + \cdot CH_2COCH_3 \xrightarrow{k_5} C_2H_5COCH_3$$

The rate of formation of $CH_3CO\cdot$ is given by the expression

(A) $k_1[CH_3COCH_3] - k_2[CH_3CO\cdot]$
(B) $k_1[CH_3COCH_3] + k_2[CH_3CO\cdot]$
(C) $k_1[CH_3COCH_3] + k_3[CH_3COCH_3][CH_3\cdot]$
(D) $k_1[CH_3COCH_3] - k_3[CH_3COCH_3][CH_3\cdot]$

Answers to Study Questions

1. B
2. B
3. A
4. A
5. B
6. D
7. B
8. D
9. B
10. B

Answers to Practice Questions

1. D
2. B
3. A
4. D
5. A
6. C
7. A
8. B
9. A
10. B
11. A

Reaction Dynamics

The concept of a reaction mechanism suggests that the observed rate of chemical reactions can be explained in terms of a series of elementary steps. We might then take this reasoning to the next level and ask, "What happens in each elementary step?" This question requires that we think about chemical reactions at the level of atoms and molecules. Consideration of kinetics at this level is referred to as the study of reaction dynamics.

At the most fundamental level, in order for a reaction to occur, the particles involved, be they molecules, atoms or ions, must encounter each other. Thus, the study of reaction dynamics begins with the study of collisions between molecules, examples of which appear in the transport properties chapter.

If we consider the matter of chemical reactions further, we quickly conclude that not all collisions between particles result in reactions. For example, under the normal conditions we encounter in everyday life, the molecules of nitrogen and oxygen in air are constantly colliding with each other but not reacting. The most important component of reaction dynamics that explains this observation is that in order for a reaction to occur the collision must have enough energy to overcome the *activation energy* for the reaction. The activation energy is characteristic of a reaction and represents the minimum amount of energy that must be available to colliding particles in order for a reaction to occur.

The activation energy of a reaction can also be related to structural changes that must occur in order to allow the transition from reactants to products. These structural implications can be considered in several ways, including the calculation of potential energy surfaces that map the energy of a system upon variation of structural variables (such as bond lengths) in the reaction system. While these variables can be manipulated in any manner, the activation energy of a process is generally associated with a specific structure called the transition state.

This chapter includes problems that review some of the theories used in these explanations such as collision theories and transition state theory. The interpretation of minimum energy pathway diagrams and potential energy surfaces and trajectories on these potential energy surfaces are also tested.

Study Questions

RD-1. Which statement is **NOT** true for the behavior of a catalyst?

   (A) A catalyst increases the equilibrium constant and final product concentrations.

   (B) A catalyst reduces the activation energy for a reaction.

   (C) A catalyst speeds up the rate of a reaction.

   (D) A catalyst increases the rates of both the forward and reverse reactions.

Knowledge Required: Definition of a catalyst.

Thinking it through: A catalyst speeds up the reaction rate by providing a lower energy pathway for the reaction. Catalysts increase the rates of reversible reactions in both directions. Comparison of the definition of a catalyst to these statements indicates that response **(A)** is not true.

RD-2. In the derivation of "transition state theory", the assumption is made that the reactants are

   (A) unstable relative to the products. **(B)** in equilibrium with the products.

   (C) in equilibrium with an activated complex. **(D)** hard sphere molecules.

Reaction Dynamics

Knowledge Required: Assumptions made in transition state theory.

Thinking it through: Transition state theory assumes that every set of reactants that cross a set point on the potential energy surface (saddle point) progress to reactants. This theory also assumes that the molecules have a Boltzmann distribution of energies. Application of these assumptions leads to the condition that the reactants are in equilibrium with an activated complex. Therefore response **(C)** is the best response.

RD-3. The entropy of activation of a reaction may be obtained from

  (A) the temperature dependence of the rate constant.

  (B) measurements of the heat of reaction.

  (C) the energy of activation of the reaction.

  (D) the activity coefficients of the reactants and products.

Knowledge Required: Relationship between the parameters in the Arrheuius expression and transition-state theory.

Thinking it through: The Arrhenius expression relates the temperature dependence of the rate constant to a pre-exponential factor A and the activation energy E_A: $k = Ae^{-E_A/k_BT}$ or $\ln k = \ln A - E_A/k_BT$. A plot of $\ln k$ versus $1/T$ is linear with an intercept equal to $\ln A$. Using transition-state theory A and E_A can be related to thermodynamic properties of the transition-state; $A = $ constant $\times T \times \exp(\Delta S^\ddagger /R)$. The activation energy can be related to the enthalpy of activation. Therefore the entropy of activation can be determined from the temperature dependence of the rate constant and response **(A)** is the correct answer. Measurements of the heat of reaction and activity coefficients relate to equilibrium properties of the reaction and not to the thermodynamics of the transition-state (activation), ruling out responses **(B)** and **(D)**.

RD-4. The figure shows the reaction profile for

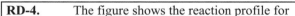

Identify the **incorrect** statement

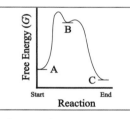

  (A) $\Delta_r G < 0$ **(B)** $\Delta G_{-1}^\ddagger > \Delta G_2^\ddagger$

  (C) $k_{-1} > k_2$ **(D)** The first transition state resembles B more closely than A

Knowledge Required: Interpreting reaction profile diagrams.

Thinking it through: By comparing G values for products, reactants and intermediates, you can determine that responses **(A)** and **(B)** are correct statements, thus incorrect responses. The first transition-state is closer in energy to B than to A and therefore response **(D)** is reasonable. Response **(C)** is an incorrect statement, thus the correct response. The reaction profile indicates that the reverse of this statement is correct, as the barrier to the first transition-state is larger than that to the second transition-state.

Reaction Dynamics

| RD-5. | For the reaction H + H–H → H–H + H, which point represents the transition state? | 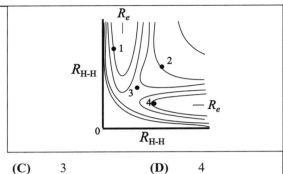 |

(A) 1 (B) 2 (C) 3 (D) 4

Knowledge Required: Identifying the transition-state on a potential energy surface.

Thinking it through: The transition-state corresponds to a maximum in the potential energy surface along the minimum energy pathway. Therefore point 3 (response **(C)**) is the correct response. Points 1 and 4 are in the product and reactant channels respectively, and point 2 is located on the repulsive wall.

RD-6. When the hydrogen and fluorine atoms

$$H_a - H_b + F \rightarrow H_a + H_b - F$$

are constrained to react in a collinear geometry, the potential energy surface for the reaction is shown. The energy contours are given in kcal·mol^{-1}. Identify the conclusion that cannot be drawn from this diagram.

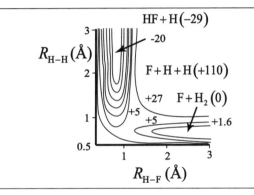

(A) The equilibrium $H_b - F$ bond distance is approximately 0.9 Å.

(B) At the saddle point, the $H_b - F$ distance is significantly longer than the equilibrium HF bond distance.

(C) The reaction is exothermic by 29 kcal·mol^{-1}.

(D) Much of the energy of the reaction is converted to translational energy.

Knowledge Required: Interpreting potential energy surfaces.

Thinking it through: Bond lengths and comparison of distances between atoms can be determined from the potential energy surface since these distances are plotted on the *x* and *y* axes. This information can be used to determine that responses **(A)** and **(B)** are correct. Response **(C)** is also a correct conclusion. The heat of reaction can be determined by comparison of the energy contours in the reactant and product channels. Response **(D)** is not a correct conclusion from the given potential energy surface. You would need to study reaction trajectories to draw a conclusion about how much energy is converted to translational energy.

RD-7. Diffusion controlled reactions have rates limited by

(A) the lifetime of the encounter pair. (B) the rate of formation of encounter pairs.

(C) activation energies. (D) diffusion of the solvent molecules.

Reaction Dynamics

Knowledge Required: Factors that contribute to reaction rates in diffusion controlled reactions.

Thinking it through: Response **(B)** is the best answer. A reaction will only occur if the reactants can encounter each other. Therefore, the rate of reaction is determined by how fast the reactants diffuse towards each other to form encounter pairs. The barrier to reaction in a diffusion controlled process is assumed to be small, so responses **(A)** and **(C)** can be eliminated. Response **(D)** is also incorrect as the solvent molecules do not react.

RD-8. For a gas phase reaction between A and B, the collision rate between A and B usually exceeds the reaction rate because the collision rate does not include the effect of the

(A) molecular mass. (B) steric factor.
(C) molecular speed. (D) cross section.

Knowledge Required: Collision rates, factors that contribute to reaction rates in gas phase reactions.

Thinking it through: The reaction rate for a gas phase reaction uses collision theory which is given by

reaction rate = (collision rate)(fraction of molecules with sufficient energy)(steric factor)

or

$$\text{rate} = Z_{AB} e^{-E_a/RT} p$$

Thus, response **(B)** is correct since the activation energy term is not included and the effects of responses **(A)**, **(C)**, and **(D)** are included in the collision rate term.

RD-9. According to absolute reaction rate (transition-state) theory, the products of a reaction are formed when

(A) the activated complex vibrates.
(B) the activated complex rotates π radians.
(C) one of the translational modes of the activated complex becomes a vibrational mode.
(D) one of the vibrational modes of the activated complex becomes a translational mode.

Knowledge Required: Concepts involved in absolute reaction rate (transition – state theory).

Thinking it through: At the transition state (the top of the potential energy surface), the activated complex has a number of vibrational modes corresponding to motions of the atoms. One of those modes (that along the reaction coordinate) has an imaginary frequency. That mode becomes a translation of the product molecules with respect to each other, thus response **(D)** is correct.

RD-10. In the absence of additional information, a lower bound for the activation energy of the endothermic reaction

$$A(g) + B(g) \rightarrow \text{Products}$$

is best approximated by

(A) the standard enthalpy change of the reaction.
(B) the larger of the bond dissociation energies of the reactants.
(C) the sum of the bond dissociation energies of the reactants.
(D) the difference of the bond dissociation energies of the reactants.

Reaction Dynamics

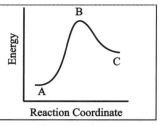

Knowledge Required: Changes in reaction coordinate diagram with ΔH, E_a.

Thinking it through: An endothermic reaction must have point A lower than point C. The lowest possible energy for the transition state (the minimum barrier at point B going from C to A is zero) is thus the same energy as C. Thus the lower bound for the forward reaction is the ΔH for the endothermic reaction, and response **(A)** is correct.

Practice Questions

1. In activated complex theory, a positive entropy of activation yields a

    (A) large Arrhenius prefactor.

    (B) small Arrhenius prefactor.

    (C) large activation energy.

    (D) small activation energy.

2. In collision theory, the steric factor is related to the

    (A) relative orientation of the colliding reactants.

    (B) atomic or molecular mass of each reactant.

    (C) highest vibrational frequency of each reactant.

    (D) molecular polarizability of each reactant.

3. Transition state theory predicts that the rate constant for a reaction is given by the expression.

    $$k = \left(\frac{k_B T}{h}\right) e^{(\Delta S^\ddagger / R)} e^{\Delta H^\ddagger / RT}$$

    Using $E_A = -\left(\frac{d \ln k}{d(1/T)}\right)$, then E_A depends on

    (A) $\Delta H^\ddagger$.   (B) $\Delta H^\ddagger$, T.

    (C) $\Delta H^\ddagger$, $\Delta S^\ddagger$, T.   (D) $\Delta S^\ddagger$, T.

4. The rate of a chemical reaction is proportional to the total number of collisions between reactant molecules per unit time. The collision rate is largest for

    (A) high collision cross section and low molecular velocity.

    (B) high collision cross section and high molecular velocity.

    (C) low collision cross section and low molecular velocity.

    (D) low collision cross section and high molecular velocity.

5. The redox reaction between Cu^+ and $Co(NH_3)_5Cl^{2+}$ has been studied in mixtures of water ($\varepsilon_{water} = 78.5$) and methanol ($\varepsilon_{methanol} = 32.6$) in the presence of an inert salt. The reaction rate is expected to

    (A) increase with added methanol and added inert salt.

    (B) decrease with added methanol and added inert salt.

    (C) increase with added methanol, but decrease with added inert salt.

    (D) remain unchanged with added methanol, but increase with added inert salt.

6. According to absolute rate theory, the pre-exponential factor in the Arrhenius equation is related to

    (A) the enthalpy difference between reactants and the activated complex.

    (B) the internal energy difference between reactants and the activated complex.

    (C) the internal energy difference between reactants and products.

    (D) the entropy difference between reactants and the activated complex.

7. A fundamental postulate of the theory of absolute reaction rates is that

   (A) the pre-exponential term is independent of temperature.
   (B) an equilibrium exists between reactants and activated complex.
   (C) the transition state contains one more degree of freedom than the reactants.
   (D) two molecules must collide.

8. Simple collision theory provides a prediction for the rates of gas phase reactions. Which statement concerning the rate equation derived from collision theory is incorrect?

   (A) Steric factors are usually included in the derivation.
   (B) The collision frequency must be divided by two.
   (C) A Boltzmann distribution of translational energies is assumed.
   (D) The pre-exponential factor obtained is proportional to $T^{1/2}$.

9. In activated complex theory, the Arrhenius pre-exponential factor can be calculated using

   (A) $\Delta S^{0\ddagger}$ (B) $\Delta H^{0\ddagger}$
   (C) $\Delta G^{0\ddagger}$ (D) E_a

10. In the Chapman mechanism for atmospheric ozone destruction
    1. $O_2 + h\nu \rightarrow 2O$
    2. $O + O_2 \rightarrow O_3$
    3. $O + O_3 \rightarrow 2O_2$
    4. $O_3 + h\nu \rightarrow O + O_2$

    the observed quantum yield can be greater than 1 because of step

    (A) 1   (B) 2   (C) 3   (D) 4

Answers to Study Questions

1. A
2. C
3. A
4. C
5. C
6. D
7. B
8. B
9. D
10. A

Answers to Practice Questions

1. A
2. A
3. B
4. B
5. A
6. D
7. B
8. A
9. A
10. A

Statistical Mechanics

The study of chemistry routinely requires the use of models at both the macroscopic and molecular levels. In some cases these two views are simply complimentary ideas that can be considered in some sense independent of each other. At other times, however, important insight is gained by relating the models at these two levels. The machinery needed to forge these relationships falls in the realm of a field called statistical mechanics.

The field of statistical mechanics provides a bridge between the macroscopic properties in traditional thermodynamics and the molecular nature of quantum mechanics. The main premise of this field is that macroscopic size samples of matter contain such vast numbers of particles that observations made at that level can be determined based on probabilistic arguments. Thus, statistical mechanics is predicated on understanding the laws of probability that explain the observations of matter.

A great deal of mathematical theory has been built up to explain probability. The key factor that allows us to apply this machinery is the realization that the behavior of molecular systems can be understood in terms of the probability distributions of the particles in quantum mechanical energy states. Partition functions are used to describe this distribution, so much of introductory statistical mechanics is based on developing this concept and its implications for molecules.

Partition functions and their applications to the calculation of thermodynamic properties such as heat capacities, entropy, etc are reviewed in this chapter. This chapter also reviews the Boltzmann distribution, relative populations, a comparison of different types of energy contributions to various thermodynamic properties, and temperature dependence of properties.

Study Questions

SM-1. Suppose two isomers, A and B, in equilibrium have the energy levels shown. Then,

(A)	A is favored at all temperatures.	**(B)**	B is favored at all temperatures.
(C)	very low temperatures favor A, very high temperatures favor B.	**(D)**	very low temperatures favor B, very high temperatures favor A.

Knowledge Required: Relationship between the population of energy levels and temperature.

Thinking it Through: Isomer A has a low and a high energy level, while isomer B has several high energy levels centered around the higher of A's levels. Therefore A will be favored at low temperatures (lower energy) because there are no corresponding low energy levels for B. B will be favored at high temperatures (high energies) because there are multiple levels available, while A has only one available high energy level. Therefore the correct response is **(C)**.

Statistical Mechanics

SM-2. Which figure **cannot** correspond to an equilibrium distribution of molecules between two quantum states at 300 K?

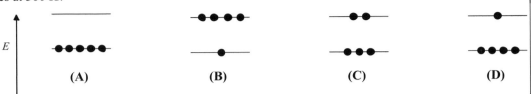

Knowledge Required: Temperature dependence of energy distributions.

Thinking it through: As temperature increases in a two-level system, the population in the lower level decreases, and the population in the upper level increases. At very high temperatures (as the temperature approaches infinity) the populations in the two levels approach equality. The maximum number of molecules allowed at the upper energy for the given system is then $5/2$, which is 2. Therefore, responses **(A)**, **(C)** and **(D)** are reasonable distributions. Response **(B)** is not reasonable, and is the correct response.

SM-3. For many molecules, only the lowest vibrational energy state is significantly populated at room temperature. In this case, the vibrational partition function is close to

(A) 0 (B) 1 (C) the temperature T (D) ∞

Knowledge Required: Temperature dependence of vibrational partition function.

Thinking it through: The vibrational partition function has the form: $\frac{1}{1-e^{-\text{constant}/T}}$. At low temperatures this function approaches 1, therefore response **(B)** is correct.

SM-4. Which partition function will be the same for gaseous H_2 and D_2 at all temperatures?

(A) translational (B) rotational
(C) vibrational (D) electronic

Knowledge Required: Mass dependence of partition functions, separation of electronic from nuclear motion.

Thinking it through: Translational, rotational, and vibrational motions correspond to the motion of the nuclei of the molecules and therefore are mass dependent. These partition functions will be different for H_2 and D_2. Electronic motion depends on the movement of electrons and therefore does not depend on mass, so the electronic partition functions will be the same for these two molecules. Response **(D)** is therefore the correct response.

SM-5. If an atom has a doubly degenerate excited state that is $\Delta\varepsilon$ cm^{-1} above the non-degenerate ground state, the electronic partition function at any temperature (T) for this atom is

(A) $1 + 2e^{-\Delta\varepsilon k_B / T}$ (B) $2 + e^{-\Delta\varepsilon / k_B T}$
(C) $1 + 2e^{-\Delta\varepsilon / k_B T}$ (D) $2 + 2e^{-\Delta\varepsilon / k_B T}$

Knowledge Required: Correct form of electronic partition function, including degneracies.

Thinking it through: The excited state is doubly degenerate, therefore a two should multiply its contribution to the total partition function and a one should multiply the ground state's contribution. This eliminates responses **(B)** and **(D)**. Response **(A)** can also be eliminated as the Boltzmann constant should be in the denominator of the exponential. Response **(C)** is therefore correct.

SM-6.	What is the molecular partition function for the two level system shown?

(A) $2e^{-\varepsilon_1/k_BT} + 3e^{-\varepsilon_2/k_BT}$

(B) $e^{-\varepsilon_1/k_BT} + e^{-\varepsilon_2/k_BT}$

(C) $e^{-2\varepsilon_1/k_BT} + e^{-3\varepsilon_2/k_BT}$

(D) $e^{-\varepsilon_1/2k_BT} + e^{-\varepsilon_2/3k_BT}$

Knowledge Required: Functional form of partition functions, including degeneracy in partition functions.

Thinking it through: The partition function is: $q = \sum g_i e^{-\varepsilon_i/k_BT}$, where g_i is the degeneracy of the ith state. Response **(A)** is the only one that includes the degeneracy correctly and is therefore the correct response.

SM-7. The order in which various molecular degrees of freedom of a gaseous diatomic molecule contribute to the heat capacity as the temperature is raised is

(A) elec > vib > rot > trans. (B) rot > vib > elec > trans.

(C) vib > elec > trans > rot. (D) trans > rot > vib > elec.

Knowledge Required: Relative importance of various molecular degrees of freedom to heat capacity. Temperature dependence of various molecular degrees of freedom.

Thinking it through: The energy available to a molecule increases as the temperature is increased. The energy required for translation < for rotational excitation < for vibrational excitation < to excite an electron, therefore the correct order is given in response **(D)**.

SM-8. The theoretical expression for the molar vibrational heat capacity of a diatomic gas is

$$C_V = \frac{R(\Theta/T)^2 e^{-\Theta/T}}{(e^{-\Theta/T} - 1)^2}$$

in which Θ is a characteristic temperature for the molecule. This equation predicts that

(A) C_V approaches zero as the temperature approaches infinity.

(B) C_V always decreases as temperature increases.

(C) C_V approaches R as the temperature approaches infinity.

(D) C_V increases without limit as the temperature approaches zero.

Knowledge Required: Understanding the temperature dependence of the vibrational heat capacity.

Thinking it through: The correct response is **(C)**. You can get to this answer by taking the limit of the above expression as T approaches infinity. This calculation is involved and would probably take too much time. Another way to approach this solution is to recall that a plot of C_V versus T/Θ approaches R as T increases; C_V approaches the classical value of R for a diatomic gas.

SM-9. Consider a system of independent particles each of which can exist in two states, A and B. The energies and degeneracies are given in the table:

State	Degeneracy	Energy / J·mol⁻¹
A	1	0
B	2	600

At approximately what temperature will the population in state A be equal to that in state B?

(A) 0 K (B) 57 K (C) 104 K (D) 273 K

Statistical Mechanics

Knowledge Required: The Boltzmann distribution for temperature dependence of states.

Thinking it through: The ratio of the populations is given by the Boltzmann distribution to be (and on substituting)

$$1 = \frac{N_B}{N_A} = \frac{g_B}{g_A}e^{-(\varepsilon_B - \varepsilon_A)/RT} = \frac{2}{1}e^{-(600\ \text{J·mol}^{-1})/(8.314\ \text{J·mol}^{-1}\cdot\text{K}^{-1})T}$$

Solving for T in this calculator allowed question gives the correct response to be 104 K or response **(C)**.

SM-10. The bond length for $H^{127}I$ is 160.916 pm and that for $H^{35}Cl$ is 127.455 pm. Which statement is true at 300 K? The spacing between the rotational states

(A) is smaller for HI, therefore HI will have a higher value of q_{rot}.

(B) is larger for HI, therefore HI will have a higher value of q_{rot}.

(C) is smaller for HCl, therefore HCl will have a higher value of q_{rot}.

(D) is not enough information to determine which molecule will have a larger value of q_{rot}.

Knowledge Required: Rotational partition function (q_{rot}) dependence on different factors.

Thinking it through: The rotation partition function, q_{rot}, for a diatomic molecule is given by

$$q_{\text{rot}} = \frac{8\pi^2 I k_B T}{h^2} = \frac{8\pi^2 \mu R^2 k_B T}{h^2} \propto \mu R^2$$

Since the reduced mass for $H^{127}I$ and $H^{35}Cl$ are both approximately 1 g·mol^{-1}, $E \propto \frac{1}{R^2}$, $q \propto R^2$, and the spacing will be smaller for HI, and q_{rot} will be larger. This corresponds to response **(A)** being correct.

Practice Questions

1. Consider the rotational contribution to the heat capacity of a gas. The rotational temperatures, Θ_r, for H_2, O_2, and CH_4 are given in the table. At 120 K, which molecule will deviate most from the classical limit?

	H_2	O_2	CH_4
Θ_r	85.4	2.06	15

   (A) All three are at the classical limit.
   (B) H_2
   (C) O_2
   (D) CH_4

2. The I_2 vibrational frequency is 6.42×10^{12} s^{-1}. What is the equilibrium ratio, N_1/N_0, of the populations of the first-excited and ground vibrational states at 300 K?

   (A) 0.036   (B) 0.36
   (C) 3.6   (D) 36

3. The diagram depicts 12 molecules in a two state system at 300 K. If three more molecules are promoted to state 1 the temperature would be

   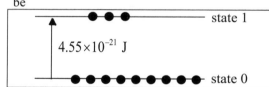

   (A) 300 K
   (B) 600 K
   (C) infinite
   (D) the temperature cannot be determined

Statistical Mechanics

4. The ground electronic state of O_2 is $^3\Sigma_g^-$. The value of the electronic partition function is

   (A) 1 because the orbital degeneracy is 1.
   (B) 2 because the "g" means "even".
   (C) 3 because the spin degeneracy is 3.
   (D) 4 because the sum of the orbital and spin degeneracy is 4.

5. In a system composed of N particles behaving classically according to Boltzmann statistics the number of particles which would be found in a state having an energy, ε, and a degeneracy, g, is directly proportional to

   (A) ε/h.
   (B) $g\varepsilon$.
   (C) $g\,e^{-\varepsilon/k_BT}$.
   (D) $g\,e^{\varepsilon/k_BT}$.

6. At high temperatures, the rotational partition limits to

   (A) 0.   (B) T.   (C) ∞.
   (D) a value which depends on the molecule.

7. As T goes to 0, the molecular partition function limits to

   (A) 0.   (B) 1.   (C) g_{elec}.   (D) ∞.

8. Above what temperature does the rotational partition function become important for H_2?

   (A) 1 K   (B) 50 K
   (C) 300 K   (D) 2000 K

9. The I_2 vibrational frequency is 6.42×10^{12} s^{-1}. The number of vibrational states with a fractional population greater than 0.2 at 300 K is

   (A) 1.   (B) 2.   (C) 5.   (D) 20.

10. The ratio of rotational partition functions for H_2 and D_2, $\dfrac{q_{H_2}}{q_{D_2}}$, at 1000 K is

    (A) 1   (B) 2   (C) $\sqrt{2}$ (D) ½

11. Which molecule has the largest value for its rotational partition function at room temperature?

    (A) $O_2(g)$ (B) $H_2(g)$
    (C) $N_2(g)$ (D) $Cl_2(g)$

Answers to Study Questions

1. C
2. B
3. B
4. D
5. C
6. A
7. D
8. C
9. C
10. A

Answers to Practice Questions

1. B
2. B
3. C
4. C
5. C
6. D
7. C
8. B
9. B
10. D
11. D

Quantum Chemistry: History and Concepts

Several fundamental concepts permeate quantum chemistry discussions and the study of the spectroscopy of atoms and molecules. The concepts dealing with the origin of spectra and the size and shape of spectral peaks are of fundamental importance. The development of Quantum Mechanics started in the late 19th and early 20th Centuries when classical physics could not explain blackbody radiation, the photoelectric effect, electron diffraction, or the luminescence spectrum of hydrogen.

Max Planck made the first breakthrough by assuming that the energy of the oscillators in a blackbody was quantized by $\varepsilon = nh\nu$ where n is an integer, ν is the frequency and h is a constant now known as Planck constant. Using this assumption the predicted spectral distribution of blackbody radiation matched the experimental distribution. Einstein applied Planck's quantization idea to the interpretation of the photoelectric effect by proposing that radiation itself existed as small packets of energy with $\varepsilon = h\nu$, called photons. For hydrogen, first Balmer and later Rydberg developed the empirical equation explaining the observed spectrum of hydrogen,

$$\tilde{\nu} = R_H \left(\frac{1}{n_1^2} - \frac{1}{n_2^2} \right),$$

where R_H is the Rydberg constant. Bohr put forward a theoretical explanation for the Rydberg equation by assuming that angular momentum of the electron in a hydrogen atom is quantized in units of $\frac{h}{2\pi} = \hbar$. de Broglie extended the Einstein equation for the momentum of a photon to any particle of matter by relating particle momentum to an associated wavelength, $\lambda = \frac{h}{p} = \frac{h}{m\nu}$, called the de Broglie wavelength.

The early 20th Century saw the development of the Schrödinger equation, a partial differential equation describing the wave properties of matter. Solutions to the Schrödinger equation are called wave functions. The Schrödinger equation and solutions to various chemically relevant systems are discussed later. However, several fundamental concepts that follow from the application of the Schrödinger equation are presented here.

The first concept is that of the energy levels and the number of electrons that can be assigned to an energy level described by a wave function. This concept is a consequence of the requirement that a wave function for fermions, of which electrons are an example, be antisymmetric. For an antisymmetric wave function an interchange of the coordinates of two electrons results in a wave function with the opposite sign, e.g.

$$\psi(1,2) = -\psi(2,1)$$

This requirement is equivalently stated as no two electrons can occupy the same spatial orbital but if they do they must have opposite spins.

The de Broglie wave-particle duality suggested that light could be used to observe an electron. This led Heisenberg to the Uncertainty Principle, $\Delta x \Delta p_x \geq \frac{1}{2}\hbar$. This statement of the Uncertainty Principle is a consequence of the mathematical statement that the operators for position and momentum do not commute. Another consequence of the particle-wave duality is the correspondence principle which says that classical and quantum mechanical results merge in the limit of high quantum numbers.

Chemists typically use Schrödinger's wave equation to create quantum mechanical descriptions of atoms and molecules. The time-independent Schrödinger equation is $\hat{H}\psi = E\psi$ where $\hat{H}$ is the Hamiltonian operator, ψ the wave function and E the energy. This equation has the form of an eigenvalue equation. Functions are eigenfunctions of an operator when the result of applying the operator to the function f is the same as multiplication of f by a constant a, $\hat{A}f = af$. The wave functions and energies of systems are the eigenfunctions and eigenvalues of the Hamiltonian operator. The Hamiltonian is an example of an operator, a rule for changing one function into another function. The Hamiltonian is formed by adding together the operators describing the kinetic and potential energies of the quantum mechanical particles.

All quantum mechanical operators, including the Hamiltonian, must be linear and Hermitian. A linear operator $\hat{A}$ satisfies $\hat{A}(c_1 f + c_2 g) = c_1 \hat{A} f + c_2 \hat{A} g$, where f and g are functions and c_1 and c_2 are constants. A Hermitian operator satisfies $\int f^* \hat{A} g \, d\tau = \int g \left[\hat{A} f \right]^* d\tau$. (This restriction leads to some convenient properties for wave functions and expectation values of operators.)

Because in molecules the electrons move more quickly than the massive nuclei, we assume that the nuclei are fixed. This neglect of nuclear motion is called the Born-Oppenheimer approximation. One consequence of the Born-Oppenheimer approximation is that we solve the Schrödinger equation for the wave functions of the electrons is solved in a stationary field of fixed nuclear coordinates. The Born-Oppenheimer approximation is often confused with the Franck-Condon (FC) principle which also uses the idea that nuclei are heavy and slow moving compared to electrons. The FC principle states that because nuclei are much more massive than electrons an electronic transition takes place in the presence of fixed nuclei.

The Schrödinger equation is a differential equation that can be solved for the particle wave function and the energy of a particle. The exact Schrödinger equation depends on the potential used in the Hamiltonian. Not every mathematical solution to the Schrödinger equation will be acceptable to a chemist. Most quantum mechanical problems have boundary conditions that must be satisfied. Boundary conditions generally lead to quantization, where the eigenfunctions and eigenvalues depend on some (typically) integer value. The quantum number is often used as a subscript on the wave functions and energies to indicate which solution is under consideration.

In addition to satisfying the boundary conditions for a given problem all acceptable wave functions must be: continuous and continuously differentiable over the appropriate range; single-valued; finite-valued; and able to be normalized. Wave functions are normalized to insure that the probability of finding the particle somewhere is exactly 1. The probability density function, $\psi^* \psi d\tau$, is commonly associated with probability of "locating" the particle. The probability of the particle being in some finite range a to b can be found using

$$\text{Probability} = \int_a^b \psi^*(x) \psi(x) d\tau$$

Eigenfunctions of linear operators form orthogonal sets so that any two different functions in the set satisfy

$$\int_{\text{all space}} \psi_i^* \psi_j d\tau = 0$$

Because all quantum mechanical operators are linear, including the Hamiltonian, the wave functions that are solutions to a given Hamiltonian form an orthogonal set. Since wave functions must also be normalized, these sets are often referred to as orthonormal sets.

Quantum mechanical operators are used with the wave function to determine the average values of experimentally observable quantities. The average value, or expectation value, of an observable corresponding to a quantum mechanical operator $\hat{A}$ in the state described by ψ_i is

$$\langle \hat{A} \rangle = \int_{\text{all space}} \psi_i^* \hat{A} \psi_i \, d\tau$$

Because $\hat{A}$ must be Hermitian, the value of this integral will be a real number as it should be if it is an experimentally measurable value. If the wave function happens to be an eigenfunction of the operator $\hat{A}$, then the only possible value that the observable can take is the corresponding eigenvalue, a_i.

$$\hat{A} \psi_i = a_i \psi_i$$

Quantum Chemistry: History and Concepts

Study Questions

HC-1. De Broglie postulated that the wavelength of a particle is inversely proportional to its momentum. The constant of proportionality is

(A) Planck's constant, h. (B) Boltzmann's constant, k_B.
(C) the speed of light, c. (D) none of the above.

Knowledge Required: DeBroglie's relationship between the momentum and wavelength of a particle, $\lambda = h/p$ where h is Planck's constant.

Thinking it Through: Examination of the de Broglie relationship shows that it only contains Planck's constant. Thus there is only one correct answer, namely, response **(A)**. Responses **(B)** and **(C)** are other fundamental constants while response **(D)** is incorrect because there is a correct response.

HC-2. Because the nuclear motions are much slower than those of the electron, the molecular Schrödinger equation for the electron motion can be solved by assuming that the nuclei are at fixed locations. This is

(A) the Born-Oppenheimer approximation. (B) the time-dependent Schrödinger equation.
(C) Russell-Saunders coupling. (D) the variation method.

Knowledge Required: The relative sizes and speeds of nuclei and electrons in a molecule. The names of various simplifying assumptions used in quantum chemistry.

Thinking it Through: The statement correctly states the fundamental idea regarding relative nuclear and electron speeds. The concept of stationary nuclei is the Born-Oppenheimer approximation; the correct response is **(A)**. The Born-Oppenheimer approximation is an assumption used to simplify the solution of the time independent Schrödinger equation for molecules; response **(B)** is incorrect. Russell-Saunders coupling is a method for determining the total angular momentum of an atom; response **(C)** is incorrect. The variation method is a technique to find solutions to the Schrödinger equation; response **(D)** is incorrect.

HC-3. According to the Heisenberg Uncertainty Principle, if the operators for two physical properties do not commute then

(A) nothing can be said about the two properties.

(B) the uncertainties in the two will be the same.

(C) both properties can be measured exactly.

(D) the product of the two uncertainties is $\geq h/4\pi$.

Knowledge Required: The postulates of quantum mechanics, especially the Heisenberg statement of the Uncertainty Principle. The relationship between commuting operators and the observables they represent.

Thinking it Through: The commutator of operators $\hat{X}$ and $\hat{Y}$ is defined as $[\hat{X},\hat{Y}] = [\hat{X}\hat{Y} - \hat{Y}\hat{X}]$. If $[\hat{X}\hat{Y}] = 0$, the operators commute and the observables X and Y can be simultaneously measured to any designated precision. If $[\hat{X}\hat{Y}] \neq 0$, then the operators do not commute. The result is a reciprocal relationship between σ_x and σ_y (the uncertainties in X and Y). The uncertainty in one observable can only approach zero as the uncertainty of the other approaches infinity. Given the statement above the only correct response is **(D)**, especially given the usual application of the uncertainty principle to position and momentum, $\sigma_x \sigma_y \geq \frac{h}{4\pi}$. Response **(A)** is incorrect because operators are defined with respect to observables. Response **(B)** is incorrect because there will be a reciprocal relationship between the two uncertainties, and response **(C)** is incorrect because of the properties of non-commuting operators discussed above.

HC-4. The requirement that wavefunctions for electrons in atoms and molecules be antisymmetric with respect to interchange of any pair of electrons is

(A) the Pauli exclusion principle. (B) the Born-Oppenheimer approximation.
(C) the Heisenberg Uncertainty principle. (D) Hund's rule.

Knowledge Required: Each of the named concepts included in the question.

Thinking it Through: In general chemistry you learned that the Pauli exclusion principle stated that no two electrons in an atom can have the same 4 quantum numbers. In quantum mechanics this statement follows a postulate that states that all electronic wavefunctions must be antisymmetric when the coordinates of two electrons are interchanged. Clearly **(A)** is the correct response. Responses **(B)**, **(C)**, and **(D)** are incorrect for the following reasons. The Born-Oppenheimer approximation refers to an assumption that permits finding the wave functions for electrons in the presence of stationary nuclei; the Heisenberg Uncertainty Principle refers to the value of the commutator for two operators representing pairs of observables, and Hund's rule refers to a method for choosing a ground electronic state of an atom in terms of the largest multiplicity and highest total angular momentum.

HC-5. For electrons emitted due to the photoelectric effect, the

(A) kinetic energy and current are functions of intensity of the incident light.

(B) kinetic energy and current are functions of frequency of the incident light.

(C) kinetic energy is a function of intensity and the current is a function of frequency of the incident light.

(D) kinetic energy is a function of frequency and the current is a function of intensity of the incident light.

Knowledge Required: Failure of classical physics in particular, the photoelectric effect. Dependences of kinetic energy and current on frequency and intensity in photoelectric effect.

Thinking it Through: In the photoelectric effect, light strikes a metal surface and electrons (photoelectrons) are released from the surface. The measureable quantities are the frequency and intensity of the incoming light and the kinetic energy and number (current) of the resulting photoelectrons. The photoelectric effect provided evidence for light behaving as a particle with $E = h\nu$ and the intensity related to the number of particles striking the surface. By conservation of energy, increasing the frequency of the incoming light will increase the energy of the light and thus increase the energy of the resulting photoelectrons, while increasing the intensity of the light (the number of light particles) will increase the number of photoelectrons released, the current will increase. Thus response **(D)** is the correct response.

Quantum Chemistry: History and Concepts

HC-6. In the Schrödinger equation the quantity, $\hat{H}$ (Hamiltonian), represents the

(A) wave function for the system (B) probability density
(C) momentum operator (D) total energy operator

Knowledge Required: The basic form of the time independent Schrödinger equation is an eigenvalue equation: $\hat{H}\psi = E\psi$, where $\hat{H}$ is the Hamiltonian operator, ψ the wave function and E the energy. Physical observables, such as the energy or the momentum, are eigenvalues of the corresponding operator.

Thinking it Through: Since the Schrödinger equation is an eigenvalue equation where the operator is $\hat{H}$ and the eigenvalue is the total energy, the Hamiltonian is effectively the total energy operator. The correct response is **(D)**.

HC-7. The total energy of a particle (mass = m) moving in the x-direction with momentum p_x is $(p_x)^2/(2m)$. The Hamiltonian operator for this system is

(A) $-\left(\dfrac{\partial^2}{\partial x^2}\right)$

(B) $\left(\dfrac{h^2}{8\pi^2 m}\right)\left(\dfrac{\partial^2}{\partial x^2}\right)$

(C) $-\left(\dfrac{\hbar^2}{2\pi^2 m}\right)\left(\dfrac{\partial}{\partial x}\right)$

(D) $-\left(\dfrac{h^2}{8\pi^2 m}\right)\left(\dfrac{\partial^2}{\partial x^2}\right)$

Knowledge Required: The conversion from classical quantities to the corresponding quantum mechanical operator; the operator corresponding to the total energy is the Hamiltonian; how to work with complex numbers; $\hbar$ is $h/2\pi$.

Thinking it Through: To find the Hamiltonian, write down the classical expression for the energy given in the problem and then replace the classical quantity p_x with the appropriate operator: $-i\hbar\dfrac{\partial}{\partial x}$.

$$E = \dfrac{p_x^2}{2m}$$

$$\hat{H} \equiv \dfrac{\hat{p}_x^2}{2m} \equiv \dfrac{-i\hbar\dfrac{\partial}{\partial x}\cdot -i\hbar\dfrac{\partial}{\partial x}}{2m}$$

It's a subtle point, but notice that $\hat{p}_x^2$ does not mean multiply the operator times itself, instead it means to apply the operator twice to the target function. Pulling out all the constants, noting that i^2 is -1 and $\hbar$ is $h/2\pi$ yields the final expression.

$$\hat{H} \equiv \dfrac{i^2\hbar^2}{2m}\dfrac{\partial}{\partial x}\cdot\dfrac{\partial}{\partial x} \equiv \dfrac{i^2\hbar^2}{2m}\dfrac{\partial^2}{\partial x^2} \equiv \dfrac{-1}{2m}\left(\dfrac{h}{2\pi}\right)^2\dfrac{\partial^2}{\partial x^2} \equiv -\left(\dfrac{h^2}{8\pi^2 m}\right)\left(\dfrac{\partial^2}{\partial x^2}\right)$$

The correct response is **(D)**.

HC-8. According to the postulates of quantum mechanics, which one of the following functions is acceptable as a wave function in the region $-\infty < x < \infty$?

(A) e^{-x} (B) $\dfrac{1}{x^2+1}$ (C) e^x (D) $|x|$

Knowledge Required: The requirements for an acceptable wave function are that it be continuous and continuously differentiable over the range considered; single-valued; finite-valued; and able to be normalized.

Thinking it Through: Both e^{-x} and e^{x} have infinite values in the range $-\infty < x < \infty$, at $x = -\infty$ and $x = +\infty$, respectively. Neither responses **(A)** nor **(C)** are thus acceptable wave functions. The function x is also not finitely valued on the range indicated and is not continuously differentiable. The function $\dfrac{1}{x^2+1}$, meets all the requirements. It is not normalized, but is normalizable. The correct response is **(B)**.

HC-9. When the hydrogen atomic $2p_x$ is acted upon by the operator for the z-component of the angular momentum, $(\hat{L}_z)$ we find that

(A) $2p_x$ is not an eigenfunction of $\hat{L}_z$, thus there is no definite value.

(B) $2p_x$ is an eigenfunction of $\hat{L}_z$, with eigenvalue $0\hbar$.

(C) $2p_x$ is an eigenfunction of $\hat{L}_z$, with eigenvalue $1\hbar$.

(D) $2p_x$ is an eigenfunction of $\hat{L}_z$, with eigenvalue $-1\hbar$.

Knowledge Required: An eigenvalue a_i of an operator $\hat{A}$ for a function f satisfies $(\hat{A}f_i = a_i f_i)$; the properties of operators. Hydrogen orbitals designated p_x and p_y are linear combinations of eigenfunctions of $\hat{L}_z$.

Thinking it Through: Applying the $\hat{L}_z$ operator to the $2p_x$ function yields

$$\hat{L}_z \psi_{2p_x} = \hat{L}_z \left(\frac{(\psi_{211} + \psi_{21-1})}{\sqrt{2}} \right)$$

$$= \frac{(\hat{L}_z \psi_{211} + \hat{L}_z \psi_{21-1})}{\sqrt{2}}$$

$$= \frac{(+1\hbar \psi_{211} + (-1\hbar) \psi_{21-1})}{\sqrt{2}}$$

$$= \hbar \frac{(\psi_{211} - \psi_{21-1})}{\sqrt{2}}$$

$$\neq \text{constant} \times \sqrt{2}(\psi_{211} + \psi_{21-1})$$

Since the result of $\hat{L}_z$ acting on $2p_x$ is not a constant multiplied by the $2p_x$ function, there is no definite eigenfunction. The correct response is **(A)**.

HC-10. The quantum mechanical operator for velocity is given by

(A) $-i\hbar \dfrac{d}{dx}$

(B) $-i\hbar \dfrac{d^2}{dx^2} + \hat{V}$

(C) $-\dfrac{\hbar^2}{i} \dfrac{d}{dv_x}$

(D) $-\dfrac{i\hbar}{m} \dfrac{d}{dx}$

Quantum Chemistry: History and Concepts

Knowledge Required: The classical expression for the velocity in terms of momentum is $v = \frac{p}{m}$. Quantum mechanical operators are formed by replacing the expression for the classical momentum with $-i\hbar \frac{d}{dx}$.

Thinking it Through: Replacing p in the expression for velocity with the appropriate quantum mechanical operator yields

$$\hat{v} = \frac{\hat{p}}{m} = \frac{-i\hbar \frac{\partial}{\partial x}}{m} = \frac{-i\hbar}{m} \frac{\partial}{\partial x}$$

Response **(D)** is correct.

Practice Questions

1. The Born-Oppenheimer approximation states that
   (A) the Schrödinger equation is solved assuming that the nuclei are stationary.
   (B) wave functions are written so that they are antisymmetric with respect to exchange of electrons.
   (C) conjugated organic molecules are approximated by considering only their π electrons.
   (D) nuclei are assumed to be stationary during an electronic transition.

2. The commutator of two operators, $\hat{\alpha}$ and $\hat{\beta}$, $[\hat{\alpha}, \hat{\beta}]$ equals 0. The physical properties A and B associated with the two operators.
   (A) are coupled together by an uncertainty relationship.
   (B) may be simultaneously determined with unlimited precision.
   (C) cannot be determined at all.
   (D) are equal.

3. To describe the wave nature of matter, de Broglie proposed that $\lambda = h/mv$. In arriving at this equation, he did **NOT** use
   (A) $E_{photon} = h\nu$
   (B) $\hat{H}\psi = E\psi$
   (C) $E_{photon} = mc^2$
   (D) $c = \lambda\nu$

4. Two wave functions, ψ_1 and ψ_2, are orthogonal if
   (A) $\int \psi_1^* \psi_1 d\tau = 1$
   (B) $\int \psi_1^* \psi_2 d\tau > 0$
   (C) $\int \psi_1^* z \psi_2 d\tau = 0$
   (D) $\int \psi_1^* \psi_2 d\tau = 0$

5. There are many mathematically acceptable solutions for the Schrödinger equation for any particular system, but only certain ones are physically acceptable. This is because
   (A) physically acceptable wave functions must be finite, single valued, and continuous; and have a continuous first derivative.
   (B) all wave functions must be complex.
   (C) all physical quantities are real.
   (D) all solutions must be eigenfunctions of the momentum operator.

6. A nodal surface is best described as a surface
   - (A) where the wave function approaches ∞.
   - (B) across which a particle does not move.
   - (C) separating regions of different sign for a wave function.
   - (D) where the kinetic energy of a particle is zero.

7. An eigenfunction of the operator, $\frac{d}{dx}$, is
   - (A) ax
   - (B) e^{ax}
   - (C) $\cos(ax)$
   - (D) e^{-ax^2}

8. Which statement is true?
   - (A) ψ is said to be *normalized* if $\int \psi d\tau = 1$, where the integration is over the complete range of all coordinates of ψ.
   - (B) The only mathematical requirement for a valid stationary-state wave function, ψ, for a system is that ψ be a solution to Schrödinger's time-independent equation for that system.
   - (C) A sum of two *degenerate* eigenfunctions for an operator is still an eigenfunction for that operator.
   - (D) ψ_1 and ψ_2 are orthogonal if $\int \psi_1 \psi_2 d\tau = 1$.

9. If $\psi(x)$ is the normalized wave function for a particle in one dimension, the average value for the momentum can be calculated using the momentum operator, $\hat{p}_x$, from the integral
   - (A) $\int \psi(x)^* \hat{p}_x \psi(x) dx$
   - (B) $\int \psi^*(x) \hat{p}_x dx$
   - (C) $\int \hat{p}_x \psi(x) dx$
   - (D) $\int \psi^*(x) \psi(x) \hat{p}_x dx$

10. Which function is an acceptable wave function over the indicated interval?
    - (A) e^{-x} over the range ∞ to –∞
    - (B) e^x over the range 0 to ∞
    - (C) e^{-x} over the range 0 to –∞
    - (D) e^{-x} over the range 0 to ∞

11. In a photoelectric effect experiment with photons of energy greater than the work function for the material, the number of photoelectrons ejected from a metallic surface depends on
    - (A) the intensity of the light hitting the metal.
    - (B) the angle at which the photons hit the metal.
    - (C) the wavelength of the light hitting the metal.
    - (D) both the intensity and wavelength of the light hitting the metal.

Answers to Study Questions

1. A
2. A
3. D
4. A
5. D
6. D
7. D
8. B
9. A
10. D

Answers to Practice Questions

1. A
2. B
3. B
4. D
5. A
6. C
7. B
8. C
9. A
10. D
11. A

Simple Analytical Quantum Mechanical Model Systems

The particle in-a-box (PIB) is perhaps the first model encountered in quantum chemistry. The 1-D model shown has L for the length of a box and infinitely high potential energy barriers holding the particle in the box. Within the box the potential energy, V, is zero. You should be familiar with the energies, wave functions, boundary conditions, and degeneracies for a number of simple model systems including the particle in the box in one and three dimensions. The PIB wave functions are orthonormal with energy levels $E_n = \dfrac{n^2 \pi^2 \hbar^2}{2mL^2} = n^2 \left(\dfrac{h^2}{8mL^2} \right)$.

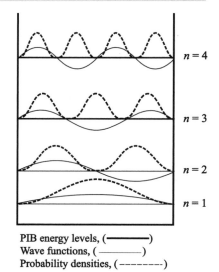

PIB energy levels, (———)
Wave functions, (———)
Probability densities, (- - - - - -)

For clarity, energies are not to scale

The PIB model is significant because we use it to understand UV-vis spectroscopy of molecules containing alternating double and single bonds. For such a system the particles are the π electrons in the conjugated part of the molecule. It is important to distinguish between an energy level and the difference between energy levels for the PIB. ΔE between a HOMO and LUMO is the energy required by a photon to cause the electron to jump from the HOMO to the LUMO.

The PIB model can be extended to boxes with various types of potential barriers. As seen for the barrier on the right, the probability distribution extends into the forbidden region, i.e., the region where the potential barrier height is higher than the energy of the particle. This is known as tunneling. Tunneling has important applications for chemistry and is used to explain point mutations in biology, certain chemical reactions requiring proton or electron transfer, and scanning tunneling microscopy.

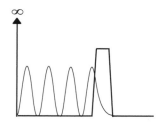

A useful model for characterizing molecular functional groups is the harmonic oscillator with $V = \frac{1}{2}kx^2$. This model provides a simple mathematical basis for understanding bond lengths, bond force constants, and infrared spectra. Again, you should be familiar with the energies, wave functions, boundary conditions and degeneracies.

The solution for the harmonic oscillator gives quantized energies with quantum numbers υ

$$E = \hbar \left(\dfrac{k}{\mu} \right)^{1/2} \left(\upsilon + \dfrac{1}{2} \right) = \hbar \omega \left(\upsilon + \dfrac{1}{2} \right) \text{ where } \upsilon = 0, 1, 2, \ldots$$

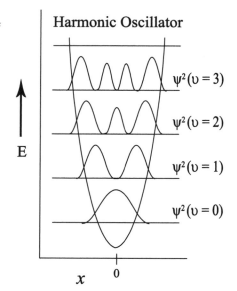

In the figure on the right we see that the energy levels are equally spaced, there is a zero point vibrational energy when $\upsilon = 0$, and E goes to infinity as υ goes to infinity. This model, based on a quadratic potential, does not account for the fact that real chemical bonds dissociate when stretched. The probability densities describe the probability of locating the oscillator at various extensions. These functions show that only certain extensions of the oscillator are very probable and others not probable. In the ground state the most probable position is the normal equilibrium distance. In the first excited state this distance is not probable at all. As υ increases the maxima of probability along the various extensions of the oscillator get closer together so that at very large values of υ the quantum mechanical oscillator behaves like a classical oscillator. An important observation is that the quantum mechanical oscillator has a small but finite probability to extend beyond the classical limit. This feature of the harmonic oscillator wave functions is also tunneling.

Adding cubic terms to the Hooke's law potential would transform it to an anharmonic potential. The Morse potential is an example of an anharmonic function because it involves an exponential function in the potential (this expansion of the exponential involves terms of many powers. Notice that the observed frequency is directly proportional to $k^{½}$ and inversely proportional to $\mu^{½}$ where k is the force constant of the bond and μ is the reduced mass. Force constants tell us how stiff a bond is.

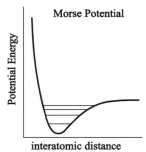

A better potential energy function for the vibration is the Morse potential. The Morse and harmonic oscillator potentials resemble each other closely at the bottom of the energy well. Solving the Schrödinger equation using the Morse potential gives energy levels that get closer together as υ increases leading to dissociation at the plateau region of the potential. The Morse potential also has a zero point energy when $\upsilon = 0$.

For a line to appear in the IR spectrum the transition dipole moment must not equal zero. This requires that the vibrating atoms in a vibrational mode have a varying dipole moment during the vibration. This results in $\Delta \upsilon = \pm 1$.

In the microwave region of the electromagnetic spectrum one can observe lines for molecules that are useful for determining bond lengths and moments of inertia arising from molecular rotation. For diatomic molecules these spectral details consist of a series of approximately equally spaced lines of varying intensities as shown. Spectra of this type are obtained for diatomic molecules that have a dipole moment. Microwave spectra allow very accurate bond lengths and structures to be determined.

For the rotation of molecules, the rigid rotor model is used. In a rigid rotor approximation the internuclear distance remains constant as the molecule rotates.

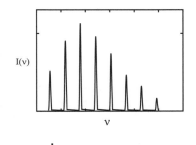

The quantized energy levels of the rigid rotor are shown to the right

$$E = J(J+1)\frac{\hbar^2}{2I}$$

and where $J \geq |m|$, where J is the angular momentum quantum number and m is the z component of J. Rotational states have $2J + 1$ degenerate energy levels. The diagram of the energy levels in units of $\frac{\hbar^2}{2I}$ shows how the energy increases with increasing J.

Only one component of angular momentum, the M_z component, can be determined in quantum mechanics. The operator for this component is

$$\hat{M}_z = -i\hbar \frac{\partial}{\partial \varphi}$$

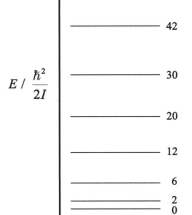

Interpretation of the J states yields some interesting insights. When $J = 0$, $E_J = 0$ meaning that the molecule is not rotating, only translation and vibration are possible. In a statistically large sample of molecules, all values of θ and φ are equally represented. When $J = 1$, the molecule is rotating. The rotation is such that only 3 planes of rotation relative to the z-axis are allowed, those corresponding to m_J values of +1, 0 and −1. Although the orientation of the plane of rotation about the z-axis is restricted, there is no restriction on the angle of φ this rotation can have. While the z component of angular momentum must have specific values, those of x and y are undetermined.

Simple Analytical Models

Allowed spectroscopic transitions require a non-zero value for the transition moment integral

$$\mu_T = \int Y_{J_f}^{m_f *} \hat{\mu} Y_{J_i}^{m_i} \sin\theta \, d\theta \, d\varphi$$

so that the dipole moment of the molecule cannot equal zero and the allowed quantum state changes are

$$\Delta J = \pm 1 \text{ and } \Delta m_J = 0, \pm 1$$

Absorption of microwave energy produces a transition between rotational energy levels given by

$$h\nu = hc\tilde{\nu} = 2(J_i + 1)\frac{\hbar^2}{2I} \text{ or } \tilde{\nu} = 2B(J_i + 1)$$

where

$$B = \frac{\hbar^2}{2hI} = \frac{h}{8\pi^2 I}$$

and J_i is the original rotational level of the transition and B is the rotational constant. From B we can determine the moment of inertia, I, and the internuclear distance using $I = \mu r^2$.

Study Questions

SAM-1.

Which sketch in the figure on the right does *not* represent a possible wave function for a one-dimensional potential energy well of finite height?

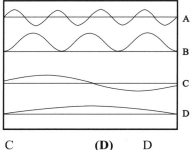

(A)   A   (B)   B   (C)   C   (D)   D

Knowledge Required: The properties of wave functions and that probability density functions that are obtained from the product of the wave function with its complex conjugate.

Thinking it Through: First look at the figure and observe that responses **(A)**, **(C)**, and **(D)** represent wave functions for the $n = 1$, $n = 2$ and $n = 4$ energy levels of a box with finite barriers. Response **(B)** does not represent a wave function; it represents a probability density. The correct response is **(B)**. When we look at A, C, and D we see that the wave functions are for energy levels increasing in order D < C < A. Response **(A)** is above the barrier height for the wall and has a higher frequency than the others. Response **(D)** has no nodes so it must be associated with the first energy level where $n = 1$. Response **(C)** has one node so it must be the wave function for energy level where $n = 2$.

SAM-2. The energy of a particle in a three dimensional box with equal sides is given by

$$E_{n_x, n_y, n_z} = (n_x^2 + n_y^2 + n_z^2)\left(\frac{h^2}{8mL^2}\right)$$ where n_x, n_y, and n_z are the quantum numbers defining an

independent state. The degeneracies of the first, second, and third energy levels are

(A)   1, 2, 3   (B)   1, 3, 1   (C)   1, 3, 3   (D)   1, 2, 2

Simple Analytical Models

Knowledge Required: The energy of the energy levels increases as the quantum numbers representing the levels increase; the permutation of the integers 1, 2, and 3 over the n_x, n_y, and n_z quantum numbers leads to degeneracy of energy levels.

Thinking it Through: Each n can vary independently and all three quantum numbers can have the same value $n_x = 1$, $n_y = 1$, and $n_z = 1$ defines the lowest energy level. There is only one unique state for the lowest energy level. The next level has one of the n values equal to 2. There are three ways to get this energy state, namely, 2, 1, 1, or 1, 2, 1, or 1, 1, 2 making this state threefold degenerate. The next highest energy state is given by the set of quantum numbers 2, 2, 1. This, too, has three possible states with the same energy, namely those described by quantum numbers 2, 2, 1, or 2, 1, 2, or 1, 2, 2. The correct response is **(C)**. Responses **(A)**, **(B)**, and **(D)** refer to higher energy levels as the sum of squares increase even more.

SAM-3. For a particle in a one-dimensional box with finite walls at $x = 0$ and $x = L$, which one of the following statements is true?

(A) The probability of finding the particle in the left-hand half of the box depends upon the quantum number n.

(B) The average value of the position of the particle is $L / 2$ for any allowed state.

(C) In the lowest energy state the probability of finding the particle in the middle third of the box is zero.

(D) The separation between energy levels increases as the mass of the particle increases.

Knowledge Required: The probability of finding a particle in the box is given by the probability distribution function which depends on the wave function squared. The functions used to describe the particle in a box energy levels and how these functions look when squared. The expression that is used to calculate the energy of a particle in a box.

Thinking it Through: Let's look at this question by eliminating wrong responses. First eliminate response **(D)** because in the equation for the ΔE we find that since mass (m) is in the denominator the spacing of energy levels decreases as m increases. Also eliminate response **(C)** because the wave function squared for the lowest energy has a maximum at the center of the box. Finally eliminate response **(A)** because the probability densities for all energy levels of a particle in a box are symmetric. The total probability for finding a particle is given by the area under the probability density. Since area on the right side of the box equals the area on the left hand side of the box and the total probability is 1, the probability on each side of the box is ½ for all values of n. Only response **(B)** is correct.

SAM-4. The most important defect of the simple harmonic oscillator (SHO) approximation for the upper vibrational levels of a molecule is that

(A) all SHO levels are evenly spaced.

(B) there is only a finite number of SHO levels.

(C) SHO levels do not change with polarizability.

(D) the SHO model only allows symmetric stretches.

Knowledge Required: The simple harmonic oscillator model is based on a simple quadratic function representation of a bond between two atoms. This model gives equally spaced energy levels, and represents the harmonic vibrations of real bonds especially well in the region of the minimum in the potential.

Thinking it Through: The solutions to the Schrödinger equation for the simple harmonic oscillator model are a series of functions identified by quantum numbers υ ranging from zero to infinity. The SHO model can be applied to any normal mode of vibration including asymmetric stretches and bending modes. The SHO does not consider bond polarizability as part of the model. Consequently only response **(A)** is correct.

SAM-5. For a harmonic oscillator, $V(x) = \tfrac{1}{2}kx^2$ and the wave function is $\psi_0(x)$. If an anharmonic energy term is given as $V'(x) = -\tfrac{1}{6}\alpha x^3$, then the energy for an anharmonic oscillator is

(A) $\tfrac{1}{2}h\nu + \int \psi_0^*(x)\left[-\tfrac{1}{6}\alpha x^3\right]\psi_0(x)dx$

(B) $\int \psi_0^*(x)\left[\tfrac{1}{6}\alpha x^3\right]\psi_0(x)dx$

(C) $\tfrac{1}{2}h\nu + \int \psi_0^*(x)\left[\tfrac{1}{2}kx^3 - \tfrac{1}{6}\alpha x^3\right]\psi_0(x)dx$

(D) $\tfrac{1}{2}h\nu + \int\left[-\tfrac{1}{6}\alpha x^3\right]dx$

Knowledge Required: One can include an anharmonic term in the Hamiltonian of the Schrödinger equation for an oscillator and then separate the parts to obtain the solutions. Including an anharmonic term is a simple example of how perturbation theory is used to obtain more useful wave functions from the Schrödinger equation.

Thinking it Through: First write the Hamiltonian for the anharmonic oscillator, $\hat{H} = \hat{H}_0 + \hat{H}_1$, where $\hat{H}_1 = -\tfrac{1}{6}\alpha x^3$. Next write the Schrödinger equation $E = \int \psi_0^*(x)\left(\hat{H}_0 - \tfrac{1}{6}\alpha x^3\right)\psi_0(x)d\tau$ and expand the integral to obtain $E = E_0 + \int \psi_0^*(x)\left(-\tfrac{1}{6}\alpha x^3\right)\psi_0(x)d\tau$ where $E_0 = \tfrac{1}{2}h\nu$. After examining the choices we see that the correct response is **(A)**. Response **(B)** neglects the E_0 term; response **(C)** includes an extraneous cubic potential term; response **(D)** is missing the wave functions in the integral.

SAM-6. In the quantum mechanical solution for the rigid rotor, the square of the angular momentum is given by

$$L^2 = J(J+1)\left(\tfrac{h}{2\pi}\right)^2$$

If $J = 2$, the possible values for the z component of the angular momentum are (in units of $\tfrac{h}{2\pi}$)

(A) 2, 1, 0, −1, −2 (B) 1, 0, −1 (C) 0 (D) 2, 1, 6

Knowledge Required: The relationship between the eigenvalues of the total angular momentum squared operator $\hat{L}^2$ and the z-component of angular momentum $\hat{L}_z$ and between quantum states described by l and m_l.

Thinking it Through: The total angular momentum squared, the eigenvalue for $\hat{L}^2$ is given by $L^2 = J(J+1)\left(\tfrac{h}{2\pi}\right)^2$. The quantum number J labels the rotational quantum state. Each quantum state has a degeneracy of $2J + 1$. The eigenvalues for the z-component of the angular momentum states ranging from $+J$ to $-J$ are called m_J. Thus when $J = 2$, $m_J = 2, 1, 0, -1, -2$. Thus the correct response for this question is **(A)**. Response **(B)** corresponds to a state with $J = 1$; response **(C)** corresponds to a state where $J = 0$. Response **(D)** has no relationship to rigid rotor quantum states.

SAM-7. In units of $\hbar^2/2I$, what is the energy separation between adjacent rotational levels J and $J+1$ of a freely rotating linear molecule?

(A) $J(J+1)$

(B) $(2J+1)J(J+1)$

(C) $\left[\tfrac{(2J+1)}{2J}\right]J(J+1)$

(D) $2(J+1)$

Knowledge Required: The relationship between quantum state energies and the peaks seen in a spectrum.

Thinking it Through: The peaks in the microwave spectrum for a diatomic molecule are due to transitions occurring between rotational energy level of a molecule. The energy for each energy level is given by $E = J(J+1)\frac{\hbar^2}{2I}$. The energy for a transition between states is given by $\Delta E = h\nu = 2(J+1)\frac{\hbar^2}{2I}$ and the allowed transitions have $\Delta J = \pm 1$. Thus the correct response is **(D)**. Response **(A)** is found as part of the angular momentum eigenvalue. Responses **(B)** and **(C)** do not have the correct form for an answer to this question.

SAM-8. The intensities and peak positions shown here were computed for one branch in the rotation-vibration spectrum of HCl molecule in the microwave region using the rigid rotor model at 298 K. When the temperature is increased to 1000 K we would expect the spectrum to change so that

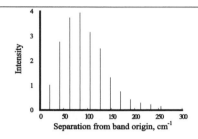

(A) the maximum intensity would occur for a peak at a higher wavenumber.

(B) the spectral lines corresponding to a given transition shift to wavenumber.

(C) the current peak with maximum intensity would remain the most intense but additional peaks at high wavenumber would appear.

(D) all peaks would retain their current relative intensities and show an increase in absolute intensity.

Knowledge Required: Ro-vibrational spectra appear in the infrared region of the electromagnetic spectrum. Such spectra consist of a series of nearly equally spaced lines, these lines vary in intensity because of consideration of the Boltzmann distribution and the degeneracy of each state.

Thinking it Through: The spectrum in the question clearly shows a series of equally spaced lines in the correct frequency range. By just considering the Boltzmann distribution the maximum would shift to a higher frequency if temperature is increased because there would be sufficient energy to populate higher quantum states. Thus the correct response is **(A)**. Responses **(B)**, **(C)**, and **(D)** all misinterpret the Boltzmann distribution law. We cannot shift the frequencies as in response **(C)** because these are determined by the difference between energy levels in the rotor. These differences are not affected by temperature. At higher temperatures the molecules have more energy so that higher energy states would become more populated at the expense of the lower energy states. Thus there would be a different energy level with a maximum population resulting in a higher frequency maximum and additional peaks observed in the spectrum at higher frequencies. Response **(D)** is wrong because to increase the absolute intensity requires increasing the number of molecules in the system. The question specifically states that only temperature is increased.

SAM-9. A portion of the ro-vibrational spectrum of HF at some temperature T is shown in the choices. The appearance of the spectrum at a high temperature could be

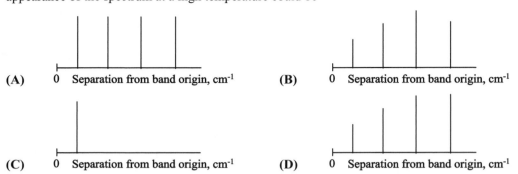

Knowledge Required: The reasons for line heights and their spacing in ro-vibrational spectra.

Thinking it Through: One branch of the infrared spectrum of a diatomic polar molecule looks like a set of nearly equally spaced lines where the lines first increase and then decrease in intensity. Knowing this you can eliminate response **(A)**. Response **(C)** shows a single line, which implies equal spacing between rotational states, which is incorrect or that only a single rotational transition can occur, also incorrect. The only viable responses are **(B)** and **(D)**. Because only four lines of the sample spectrum are shown we cannot see the decrease in intensity and increasing separation between lines that occurs at higher frequencies. Therefore, response **(D)** corresponds to the higher temperature as the decrease in intensity for the high J lines is not seen and is the correct response.

SAM-10. In the rotational spectrum of DCl, the rotational line spacings compared to those of HCl are approximately

(A) Halved. (B) Doubled. (C) Tripled. (D) Unchanged.

Knowledge Required: The lines in a pure rotational spectrum are equally spaced using the rigid rotor model and that the separation between lines equals $2B$ where $B = \dfrac{h}{8\pi^2 cI}$ and the moment of inertia $I = \mu r^2$ with μ being the reduced mass and r the internuclear distance.

Thinking it Through: $B \propto 1/I$ and thus $B \propto 1/\mu$. First evaluate μ for both HCl and DCl. The result is $\mu_{HCl} \approx 1$ and $\mu_{DCl} \approx 2$. Thus $B_{DCl}/B_{HCl} \approx 1/2$ or $B_{DCl} \propto (1/2)B_{HCl}$. The correct response is then **(A)**. Response **(B)** with ratio, B_{HCl}/B_{DCl}, is the inverse of the ratio called for in the question. Response **(C)** is not correct because there is no tripling seen in the ratios obtained from the equations for B and I. Response **(D)** is also incorrect because the reduced masses of HCl and DCl differ by a factor of 2.

Practice Questions

1. If the box length for a particle in a one-dimensional box is doubled the energy will be

   (A) 2 times the original energy.
   (B) 4 times the original energy.
   (C) ½ of the original energy.
   (D) ¼ of the original energy.

2. A particle confined in a one-dimensional box has a lowest energy of E_1. If the mass of the particle is doubled, the lowest energy will be

   (A) zero.　　(B) doubled.
   (C) $E_1/2$.　　(D) $E_1/4$.

3. The one-dimensional quantum mechanical descriptions of the harmonic oscillator and the particle-in-a-box with infinite walls *differ* in that

   (A) the energy levels for the particle-in-a-box are nondegenerate whereas those for the harmonic oscillator are degenerate.
   (B) the particle-in-a-box energy levels get farther apart as the quantum number increases, whereas the harmonic oscillator levels get closer together.
   (C) the particle-in-a-box wave functions are orthogonal to each other whereas the harmonic oscillator wave functions are not.
   (D) the particle-in-the-box wave functions do not penetrate the box walls, whereas the harmonic oscillator wave functions penetrate into the classically forbidden regions.

4. We can model 1,3,5–hexatriene as a box of length 0.904 nm, with 6 electrons in the first three energy levels. What is the frequency of an electronic transition from the highest-occupied orbital to the lowest-unoccupied orbital?

   (A) 7.78×10^{14} Hz　　(B) 1.00×10^{15} Hz
   (C) 1.24×10^{15} Hz　　(D) 1.11×10^{14} Hz

5. Which is not spectroscopic evidence for the anharmonicity of the potential energy as a function of internuclear distance?

   (A) unequal spacing of the rotational fine structure of infrared absorption bands
   (B) the existence of overtone bands in the infrared absorption spectrum
   (C) unequal spacing of the vibrational fine structure of UV–visible absorption bands
   (D) observation of combination bands

6. Compare the force constants for the bond strength in HCl and DCl.

   (A) $k(\text{HCl}) \gg k(\text{DCl})$
   (B) $k(\text{HCl}) \ll k(\text{DCl})$
   (C) $k(\text{HCl}) \approx k(\text{DCl})$
   (D) $2k(\text{HCl}) = k(\text{DCl})$

7. Given the figure which describes a Morse potential for a diatomic molecule, identify and describe region 3

   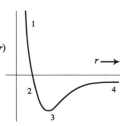

   (A) This is the attractive region of the interaction where the molecule is most stable.
   (B) This is the area describing where a molecule will dissociate because of repulsive forces.
   (C) This region is for a bound state of the molecule, but repulsive forces are pushing the atoms apart.
   (D) This region is indicative of a highly activated molecule, experiencing attractive forces, but only weakly bound.

8. For a diatomic molecule considered as a rigid rotor, $E = \left(\dfrac{\hbar^2}{2I}\right) J(J+1)$. When $J = 1$,

   (A) the angular momentum vector has only one possible orientation.
   (B) the molecule is in the lowest energy state.
   (C) the magnitude of the angular momentum is $\hbar (2)^{1/2}$.
   (D) the magnitude of the angular momentum depends on the value of m_J.

9. Which system does **NOT** have a zero point energy?

   (A) particle in a one dimensional box
   (B) one dimensional harmonic oscillator
   (C) two particle rigid rotor
   (D) hydrogen atom

10. C_{60} has 60 electrons from p orbitals not involved in the sigma bonding. If you model the electrons as particles on a sphere, what value of l is that for the highest occupied molecular orbital?

    (A) 5   (B) 6   (C) 30   (D) 60

Answers to Study Questions

1. B
2. C
3. B
4. A
5. A
6. A
7. D
8. A
9. D
10. A

Answers to Practice Questions

1. D
2. C
3. D
4. A
5. A
6. C
7. A
8. C
9. C
10. A

Model Quantum Mechanical Problems: Atomic Systems

From the perspective of a chemist, the hydrogen atom is perhaps the most important model problem in quantum mechanics. One-electron systems are the only atomic or molecular problems for which the electronic Schrödinger equation can be solved exactly. The hydrogen atom wave functions which are solutions to the Schrödinger equation provide a mathematical basis for approximate numerical solutions for multi-electron atoms and molecules. Finally, the hydrogen atom wave functions are a powerful language chemists can use to qualitatively describe the electrons in atoms and molecules.

The time-independent, non-relativistic Hamiltonian for a one-electron hydrogen-like atom is given by

$$\hat{H} = -\frac{\hbar^2}{2(m_N + m_e)}\nabla_R^2 - \frac{\hbar^2}{2\mu}\nabla^2 - \frac{Ze^2}{4\pi\varepsilon_0 r}$$

where m_N is the mass of the nucleus, m_e is the mass of the electron, ∇^2 is the Laplacian, R is a vector giving the coordinates of the center of mass of the atom, μ is the reduced mass of the electron and the nucleus, r is the distance between the nucleus and the electron, Z is the charge on the atom, e is the charge on one electron, and ε_0 is a constant called the permittivity of vacuum. The first term in the Hamiltonian describes the kinetic energy of the entire atom moving through space, the second term describes the kinetic energy of the electron moving within the atom and the third term describes the potential energy of the attraction between the nucleus and the electron.

The details of the hydrogen atom solution are not shown here. Ultimately, the eigenfunctions consist of a radial component, normally designated R_{nl}, and an angular portion designated as $Y_{lm}(\theta,\varphi)$, and given by the spherical harmonics defined for the rigid rotor. These functions introduce the three characteristic quantum numbers; n, l, and m_l. In a mathematical sense these quantum numbers designate specific functions solving the eigenvalue problem for the hydrogen atom. These functions will be given in your textbook and on the exam if needed.

Notice that two of the three p functions are not real valued ($m_l = \pm 1$). Chemists often use real valued versions of the p orbitals constructed by taking the linear combinations of p_1 and p_{-1}. This can be done since any linear combination of degenerate solutions is itself a solution with the same energy. The d orbitals can be treated similarly.

The wave functions, or the corresponding probability density functions $\left(\psi_{nlm}^* \psi_{nlm}\right)$ are sometimes graphed as surfaces of equal electron density, giving the familiar orbital shapes. The phase of the wave function (negative or positive) is sometimes indicated on the electron density surfaces as well. The energies of the hydrogen-like atom depend only on the principal quantum number, n, and are given by

$$E_n = \frac{-Z^2\mu e^4}{32\hbar^2\pi^2\varepsilon_0^2} \cdot \frac{1}{n^2}$$

Note the energies are the same as those obtained from the Bohr model.

Though the hydrogen-like atom model does not explicitly include electron spin, it can be introduced by assuming that the spin Hamiltonian can be separated out so that the spin part of the wave function can just be multiplied by the solutions to the electronic Schrödinger equation. The complete wave function for a one-electron atom including spin thus has the form

$$\psi_{nlm}(r,\theta,\varphi)\alpha \text{ or } \psi_{nlm}(r,\theta,\varphi)\beta$$

There is a spin angular momentum quantum number m_s associated with each spin function, $+\frac{1}{2}$ for α and $-\frac{1}{2}$ for β.

Model Quantum Mechanical Problems: Atomic Systems

Many electron atoms are significantly more complicated systems from the perspective of quantum mechanics. The mathematics becomes sufficiently complex that analytical solutions are no longer possible. Instead a series of approximations are introduced. Understanding the quantum nature of many-electron atoms ultimately requires the understanding of these various approximations.

The many electron wave function is generally written as a product of one-electron wave functions. It is possible to use solutions to the hydrogen atom problem in this role. Commonly, the one-electron functions are written as Slater orbitals of the form,

$$\psi(x_1, x_2, \ldots, x_N) = (N!)^{-1/2} \begin{vmatrix} \chi_i(x_1) & \chi_j(x_1) & \cdots & \chi_k(x_1) \\ \chi_i(x_2) & \chi_j(x_2) & \cdots & \chi_k(x_2) \\ \vdots & \vdots & & \vdots \\ \chi_i(x_N) & \chi_j(x_N) & \cdots & \chi_k(x_N) \end{vmatrix}$$

One feature of Slater orbitals is that they are antisymmetric with respect to the exchange of any two electrons.

Multi-electron atomic energies depend not only on the principal quantum number, n, but also on the angular quantum number, l. The familiar pattern of orbital energies and electron configurations are as depicted in the figure.

In addition to this change, things like orbital occupancy, and splitting by external magnetic fields can be addressed. The spin-orbit coupling, arising from the interaction of the electron's magnetic moment and its orbital motion about the nucleus, can be dealt with using LS coupling or Russell-Saunders coupling methods. This form of coupling can be observed in the fine structure of atomic spectroscopy. These states can be split further by the application of an external magnetic field via the Zeeman effect. Thus, as degeneracies are removed, an electron configuration, such as np^2, can be divided into terms that reflect the orbital occupancies, which can be split into multiplets by the spin-orbit interaction and these multiplets can be split into states by an applied magnetic field.

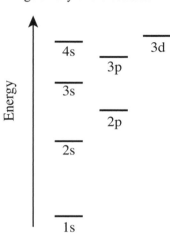

Study Questions

AS-1. Of the five orbitals in the d subshell of a hydrogen-like atom, only the angular wave function for the $m = 0$ (d_{z^2}) orbital is real. How are the other four real orbitals (d_{xy}, d_{xz}, d_{yz}, and $d_{x^2-y^2}$) generated?

(A) Using the variation method

(B) Including the spin wavefunctions

(C) Applying the Heisenberg uncertainty principle

(D) Taking linear combinations of the imaginary functions

Knowledge Required: The *d* orbitals in a hydrogen-like atom are degenerate, that is, they all have the same energy. Any linear combination of degenerate solutions to the Schrödinger equation is itself a solution with the same energy as the original solutions.

Thinking it Through: The variation method is used to generate approximate solutions to the Schrödinger equation, but the solutions for hydrogen-like atoms are exact, so we can discard response **(A)**. The spin wave functions are not functions in real Cartesian space, so multiplying wave functions by spin functions will not change the Cartesian part of the function in any way, so response **(B)** cannot be correct. The Heisenberg uncertainty principle addresses simultaneous measurements of properties, not the values of wave functions, so response **(C)** is not correct. Since any linear combination of the degenerate *d* solutions is also a solution, we can construct "to order" a set of functions that are real valued and have the same energy. Response **(D)** is correct.

AS-2. The 2s radial wave functions, R_{2s}, can be depicted in a variety of ways. Given the figure, the *y*-axis identification is

(A) R (B) R^2 (C) r^2R (D) r^2R^2

Knowledge Required: Be able to accurately sketch the wave functions for a hydrogen-like atom in several ways, including the radial distribution functions, the radial part of the solutions, and the angular part of the solutions. A radial node is a region (point, line, plane, surface) where the wave function is zero. The number of nodes in a hydrogen-like orbital is one less than the principal quantum number.

Thinking it Through: The graph does not have a node at $r = 0$. Graphs of either r^2R or r^2R^2 would have a node at $r = 0$, therefore neither response **(C)** nor **(D)** is correct. A graph of R^2 would have only positive values, therefore response **(B)** is not correct. The radial function for a 2s orbital has one node, so response **(A)** is correct.

AS-3. If a hydrogen atom with a 1s configuration is placed in a magnetic field, the electron spin degeneracy will be removed and the number of states with different energies will be

(A) one (B) two (C) three (D) four

Knowledge required: The term degenerate refers to a set of states described by different quantum numbers but having the same energy. The orbital degeneracy of any hydrogen atom energy level depends on the principal quantum number, n. The electron spin degeneracy of any given orbital is two, corresponding to spin angular momentum quantum numbers of $+\frac{1}{2}$ and $-\frac{1}{2}$.

Thinking it through: The orbital degeneracy of the $n = 1$ configuration is 1. This orbital has two possible spin states, which are ordinarily degenerate. In a magnetic field, the spin states are no longer degenerate, so they would have different energies. The correct response is therefore **(B)**.

AS-4. A certain hydrogen-like atom (with only one electron) emits light with a frequency of 2.2×10^{16} Hz associated with the $n = 2$ to $n = 1$ transition. What frequency light will be emitted for the $n = 3$ to $n = 2$ transition?

(A) 3.08×10^{15} Hz (B) 3.70×10^{15} Hz

(C) 4.11×10^{15} Hz (D) 7.40×10^{15} Hz

Model Quantum Mechanical Problems: Atomic Systems

Knowledge Required: The frequency of light emitted as a result of a transition between two states is directly proportional to the energy difference between the states. The energy of a state in a hydrogen-like atom (any atom with just one electron) is inversely proportional to the square of the principal quantum number n of the state.

Thinking it Through: Because the energy of a state is inversely proportional to the principal quantum number, the states will be more closely spaced in energy as n increases. Thus, the energy difference between states $n = 3$ and $n = 2$ will be less than between states $n = 2$ and $n = 1$, and the frequency of light emitted will be smaller. Using the relationship between principal quantum number and energy, write

$$\Delta E(n_2 \to n_1) = E_{n_2} - E_{n_1}$$
$$\propto \frac{1}{n_2^2} - \frac{1}{n_1^2}$$
$$= \text{constant} \times \left(\frac{1}{n_2^2} - \frac{1}{n_1^2}\right)$$

Because $\nu \propto \Delta E$, the ratio of the frequencies is the same as the ratio of the energy differences. Using the relationship between the energy differences and the principal quantum numbers yields

$$\frac{\nu_{3\to 2}}{\nu_{2\to 1}} = \frac{\text{constant} \times \left(\frac{1}{3^2} - \frac{1}{2^2}\right)}{\text{constant} \times \left(\frac{1}{2^2} - \frac{1}{1^2}\right)} \qquad \nu_{3\to 2} = \frac{5}{27}\nu_{2\to 1}$$

The frequency of the $n = 3$ to $n = 2$ transition thus should roughly be $\frac{1}{5}$ of the $n = 2$ to $n = 1$ transition, response **(C)** is correct.

AS-5. The square of the wave function, ψ_{1s}^2, for the hydrogen 1s orbital has

(A) A maximum at $r = 0$ (B) A minimum at $r = 0$
(C) A maximum at $r = a_0$ (D) A minimum at $r = a_0$

Knowledge Required: Graphical representation of hydrogen 1s orbital and densities.

Thinking It Through: The square of the hydrogen 1s wave function decreases exponentially from $r = 0$ to $r = \infty$. There is a maximum at $r = 0$, thus response **(A)** is the correct response. Response **(B)** would be correct for $r^2\psi_{1s}^2$.

AS-6. Which statement about the hydrogen atom is true?

(A) The energy of a state does not depend on the azimuthal quantum number.
(B) Energy levels become more widely separated as the principal quantum number increases.
(C) The total number of nodes in a wavefunction is equal to twice the principal quantum number.
(D) The $3d_{xy}$ orbital has one angular node and one radial node.

Knowledge Required: Facts about the hydrogen atom.

Thinking It Through: Response **(A)** is the correct response; the energy only depends on the principal quantum number (n) and not on the angular momentum (l) or azimuthal (m_l) quantum number. The energy levels of the hydrogen atom become closer together as n increases, therefore response **(B)** is incorrect. The total number of nodes is n not $2n$, response **(C)** is incorrect. The $3d_{xy}$ orbital has zero radial nodes and two angular nodes; response **(D)** is incorrect.

Model Quantum Mechanical Problems: Atomic Systems

AS-7. Consider the full Hamiltonian for the hydrogen atom
$$\hat{H} = -\frac{\hbar^2}{2m_{nucleus}}\nabla^2_{nucleus} - \frac{\hbar^2}{2m_{electron}}\nabla^2_{electron} - \frac{Ze^2}{4\pi\varepsilon_0 r}$$

After the coordinate transformation, the electronic kinetic energy term becomes

(A) $\quad -\dfrac{\hbar^2}{m_{nucleus} m_{electron}}\nabla^2_{nucleus}$

(B) $\quad -\dfrac{\hbar^2}{\mu}\nabla^2_\mu$

(C) $\quad -\dfrac{Ze^2}{r}$

(D) None of the above

Knowledge Required: Mathematical solution of the hydrogen atom, Schrödinger equation, coordinate transformations.

Thinking It Through: The Hamiltonian for the hydrogen atom is correctly expressed in the stem of the problem. Following the coordinate transformation the Hamiltonian becomes
$$\hat{H} = -\frac{\hbar^2}{2M}\nabla^2_M - \frac{\hbar^2}{2\mu}\nabla^2_\mu - \frac{Ze^2}{4\pi\varepsilon_0 r}$$

The first term corresponds to the kinetic energy of the whole atom; the second term to the kinetic energy relative to the nucleus, and the third term to the potential energy where M is the sum of the masses (nucleus + electron) and μ is the nucleus/electron reduced mass. Thus the correct response is response **(B)**.

AS-8. How would you calculate the probability of finding the 1s electron for a hydrogen atom at $r > a_0$?

(A) $\quad \displaystyle\int_{-\infty}^{\infty} R^*_{1s}(r) R_{1s}(r) r^2 dr$

(B) $\quad \displaystyle\int_{a_0}^{\infty} R^*_{1s}(r) r R_{1s}(r) dr$

(C) $\quad \displaystyle\int_{-\infty}^{\infty} R^*_{1s}(r) R_{1s}(r) dr$

(D) $\quad \displaystyle\int_{a_0}^{\infty} R^*_{1s}(r) R_{1s}(r) r^2 dr$

Knowledge Required: Calculations of properties.

Thinking It Through: The probability is associated with an integral of $\psi^*\psi$ (appropriate volume element) over the proper range. In the case the range for r is $a_0 \leq r \leq \infty$; thus responses **(A)** and **(C)** are incorrect responses. To separate response **(B)** from **(D)**, we look for the appropriate volume element which is $r^2 dr$ for the spherical coordinate system of the hydrogen atom problem. Thus response **(D)** is the correct response.

AS-9. The wave function for the lowest excited state of the helium atom is proportional to

(A) $1s(1)\, 1s(2) \{\alpha(1)\beta(2) - \beta(1)\alpha(2)\}$

(B) $\{1s(1)\, 2s(2) - 2s(1)\, 1s(2)\} \{\alpha(1)\beta(2) - \beta(1)\alpha(2)\}$

(C) $\{1s(1)\, 2s(2) + 2s(1)\, 1s(2)\} \{\alpha(1)\beta(2) - \beta(1)\alpha(2)\}$

(D) $\{1s(1)\, 2s(2) - 2s(1)\, 1s(2)\} \{\alpha(1)\beta(2) + \beta(1)\alpha(2)\}$

Knowledge Required: Form of the wave function for multi-electron systems. Symmetry properties of wave functions.

Thinking It Through: Response **(A)** is a correct wave function in that it is antisymmetric; however both electrons are in the 1s orbital so it is not an excited state; response **(A)** is incorrect. Response **(B)** is not a correct antisymmetric wave function so response **(B)** is incorrect. Response **(C)** and **(D)** both have the correct antisymmetric properties. Response **(C)** would be the wave function for the lowest excited singlet state having the antisymmetric spin functions while response **(D)** would be that for the lowest excited triplet state having the symmetric spin function. Since the triplet is lower, response **(D)** is correct.

AS-10. The splitting between the two lines in the yellow for the Na atom is due to

(A) splitting between $3p_{+1}$ and $3p_0$.
(B) spin-orbit coupling.
(C) splitting between $3p$ and $3d$.
(D) relativistic effects.

Knowledge Required: Splitting mechanisms for states.

Thinking It Through: The two lines for atomic sodium are close together (586 and 589 nm). Both transitions arise from a transition from the $3s^1$ ground state configuration. The two spin-orbit states resulting from the $3p^1$ configuration, $^2P_{3/2}$ and $^2P_{1/2}$, are closely spaced; response **(B)** is correct. Response **(A)** is incorrect because the m_l quantum number is not a good quantum number for multi-electron systems. Response **(C)** is incorrect because the splitting is too large, and because no $3s \rightarrow 3d$ transition is allowed. Response **(D)** is nonsense as relativity is not a factor for atoms as light as sodium.

Practice Questions

1. What is the expectation value of r, $\langle \hat{r} \rangle$, for the electron in the 1s orbital of hydrogen, where the wave function is $\psi = \frac{1}{\sqrt{\pi}} \left(\frac{Z}{a_0} \right)^{\frac{3}{2}} e^{-Zr/2a_0}$?

   (A) $\int_0^\infty \psi^* r \psi \, dr$

   (B) $\int_{-\infty}^\infty \psi^* r \psi \, dr$

   (C) $\int_0^\infty \psi^* \psi \, dr$

   (D) $\int_0^{2\pi} \int_0^\pi \int_0^\infty \psi^* r \psi r^2 \sin\theta \, dr \, d\theta \, d\varphi$

2. In the absence of a magnetic field, the order of the hydrogen atom energy levels is

   (A) $E_{1s} > E_{2s} > E_{2p} > E_{3s}$.
   (B) $E_{1s} < E_{2s} = E_{2p} < E_{3s}$.
   (C) $E_{1s} < E_{2s} < E_{2p} < E_{3s}$.
   (D) $E_{1s} > E_{2s} = E_{2p} > E_{3s}$.

3. In the presence of a magnetic field in the z direction, the ordering of the energy states of an atom is

   (A) $E_{p_{+1}} > E_{p_0} > E_{p_{-1}}$.
   (B) $E_{p_{+1}} < E_{p_0} < E_{p_{-1}}$.
   (C) $E_{p_{+1}} = E_{p_0} = E_{p_{-1}}$.
   (D) indeterminate.

4. Measurement of the z component of angular momentum for an electron in the $2p_x$ state will yield

   (A) $+1 \, \hbar$.
   (B) $-1 \, \hbar$.
   (C) $0 \, \hbar$.
   (D) $\pm 1 \, \hbar$.

5. Which graph would be best used to explain why the average electron-nucleus distance for the 1s state of the H atom is 0.529 Å?

   (A) R against r
   (B) R^2 against r
   (C) $r^2 R^2$ against r
   (D) ψ^2 against r

6. The average value of the angle φ for an electron in a H atom 1s orbital is

   (A) 0. (B) π/2. (C) π. (D) 2π.

7. The Na atom has 2 lines in the yellow region of the spectrum. These arise from transitions between which energy levels?

   (A) $3s \to 3p$
   (B) $3s \to 4s$
   (C) $3s \to 3d$
   (D) $3s \to 4p$

8. The Hg atom has a long-lived excited state which emits at 253.7 nm. The likely transition causing this emission is

   (A) $6s7s5d^{10}\ ^1S_0 \to 6s^25d^{10}\ ^1S_0$.
   (B) $6s7s5d^{10}\ ^3S_0 \to 6s^25d^{10}\ ^1S_0$.
   (C) $6s6p5d^{10}\ ^3P_1 \to 6s^25d^{10}\ ^1S_0$.
   (D) $6s6p5d^{10}\ ^1P_1 \to 6s^25d^{10}\ ^1S_0$.

9. What factor in the quantum mechanical equations of multi-electron atoms makes them so they cannot be solved analytically?

   (A) the spin operator
   (B) the electron-electron repulsion term
   (C) the need for the Born-Oppenheimer approximation
   (D) the non-zero commutator between the electron position and momentum

10. How many nodes are present in a 4d orbital of a hydrogen atom?

    (A) 3 angular nodes
    (B) 2 angular nodes and 1 radial node
    (C) 1 angular node and 2 radial nodes
    (D) 3 radial nodes

Answers to Study Questions

1. D
2. A
3. B
4. C
5. A
6. A
7. B
8. D
9. D
10. B

Answers to Practice Questions

1. D
2. B
3. A
4. D
5. C
6. C
7. A
8. C
9. B
10. B

Symmetry

Symmetry considerations allow chemists to organize information about the complex shapes of molecules and orbitals in a compact form. Information about the symmetry of orbitals and molecules can also help predict which spectral transitions will be observed. Finally, symmetry considerations can simplify calculation of the molecular wave function. One way to classify the shape of a molecule is to determine its point group by ascertaining which symmetry operations can be applied to the molecule. The symmetry elements found in molecular point (or symmetry) groups are C_n axes (an n-fold rotation axis); σ_v ("vertical" plane of symmetry that contains the principal rotational symmetry axis); σ_h ("horizontal" plane of symmetry that is perpendicular to the principal rotational symmetry axis); i (inversion through a molecular center); S_n (improper rotational axis equivalent to an n-fold rotation around the axis followed by a reflection through a plane perpendicular to the axis). A flow chart for determining the symmetry group of a molecule is typically found in your textbook.

Because the molecular symmetry operations commute with the Hamiltonian, the wave function must also be an eigenfunction of the symmetry operators. One consequence of this is that molecular orbitals will be either symmetric or anti-symmetric with respect to each symmetry element of the molecule. Molecular normal vibrational modes are also characterized using symmetry operations. Instead of symmetry elements, the irreducible representations are used. The character tables for each point group list the irreducible representations for that group. Normal modes with irreducible representations that transform as the electric dipole does (typically indicated by x, y, z in the far right hand column of a character table are IR active. Modes which transform as the polarizability does (indicated by $x^2, y^2, z^2, xy, xz, yz$) are Raman active. To determine which irreducible representation a given normal vibrational mode belongs to, apply each symmetry operation of the molecule's point group to the vibration and note whether it transforms symmetrically (no change in direction) yielding a character of 1 or asymmetrically (changes direction under the operation) with a character of -1. Match the list of characters so generated with those of one of the irreducible representations in the table. The number of normal modes is $3N-5$ for linear molecules and $3N-6$ for non-linear molecules, where N is the number of atoms in the molecule.

Atomic and molecular terms symbols provide a summary of the possible detailed electronic configurations and energies as well as a quick guide to allowed transitions. In general a term symbol has the form $^{2S+1}X_J$, where $2S+1$ is called the spin multiplicity. S is the sum of the spin angular momentum of each electron. Note that operationally only open subshells need to be considered in computing L or S. X is a letter indicating the maximum value of the total orbital angular momentum L. For $L = 0, X = $ S; $L = 1, X = $ P; $L = 2, X = $ D; etc. The possible values of L for an electron configuration range between $|l_1 + l_2| \geq L \geq |l_1 - l_2|$.

When writing electron configurations of atoms correctly one also uses Hund's maximum multiplicity rule which states that an atom in its ground state adopts a configuration with the greatest number of unpaired electrons. This generality arises out of the more specific Hund's rules which assign relative energy to spectroscopic states based on S, L, and J values. Summarizing these, the state with the largest S is most stable and when there are identical S values for two or more states then the largest L is used to determine the most stable state. Finally for states with the same S and L, the lowest energy state is chosen using the smallest J when the shell is less than half filled or the largest J for a shell that is more than half-filled.

Transitions between atomic states depend on the selection rules for electronic transitions. For many-electron atoms the electronic selection rules are

$$\Delta J = 0, \pm 1 \, (J = 0 \rightarrow J = 0 \text{ is forbidden})$$
$$\Delta L = 0, \pm 1$$
$$\Delta S = 0$$
$$\Delta l = \pm 1$$

Study Questions

S-1. What are the possible values of the term symbols for the $1s^1 2s^1$ configuration of helium?

(A) $^1S, {}^3S$ (B) ^2S only (C) $^1S, {}^3P$ (D) ^1S only

Knowledge Required: How to determine possible term symbols from electron configurations.

Thinking it Through: To determine whether a given electron configuration corresponds to an S and/or P term, find the value of L by summing the l values for the unfilled shells. In this case $L = l_{1s} + l_{2s} = 0 + 0 = 0$; $L = 0$ states are designated S. If both spins are paired $S = 0$ and the spin multiplicity is 1, alternatively the spins could be unpaired, in which case $S = 1$ and the spin multiplicity is 3. The two possible terms are thus ^1S and ^3S; **(A)** is the correct response.

S-2. What is the symmetry of BF$_3$?

(A) C_3 (B) C_{3v} (C) C_6 (D) D_{3h}

Knowledge Required: Be able to classify the symmetry of the molecule; the orbitals of a molecule must have the same symmetry as the molecule

Thinking it Through: BF$_3$ has a 3-fold axis of symmetry perpendicular to the molecular plane, a plane of symmetry (the molecular plane), 3 2-fold axes of symmetry perpendicular to the 3-fold axis, therefore it has D_{3h} symmetry. The molecule must have the same symmetry, so response **(D)** is correct.

S-3.

Using the orientation shown, choose the irreducible representation representing the vibrational mode in the figure

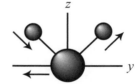

C_{2v} representation	E	C_2	σ_v	σ_v'	
A_1	1	1	1	1	z, x^2, y^2, z^2
A_2	1	1	−1	−1	xy
B_1	1	−1	1	−1	x, xz
B_2	1	−1	−1	1	y, yz

(A) A_1 (B) A_2 (C) B_1 (D) B_2

Knowledge Required: The effect of symmetry operations; the convention that the σ_v' plane is the molecular plane (This can be determined from the character table, which notes that yz transforms symmetrically with respect to σ_v').

Thinking it Through: Applying E, the identity element yields a character of 1, which cannot help discriminate between irreducible representations. Applying C_2 reverses the motions, so the character is −1, and the representation must be either B_1 or B_2. The motions are unchanged by reflection in the molecular plane (σ_v'), so σ_v' has a character of 1. The irreducible representation of this vibration is B_2, the correct response is **(D)**.

S-4. A C_{2v} molecule is in an initial state characterized by B_1 symmetry. What is the symmetry of the final state when an electric dipole transition in the z-direction is allowed?

(A) A_1 (B) A_2 (C) B_1 (D) B_2

Symmetry

Knowledge Required: Using the character table for a symmetry group to understand spectroscopic transitions.

Thinking It Through: An electron dipole transition can occur when the integral $\int \psi_f \hat{z} \psi_i \, d\tau$ is non-zero. The integral can be shown to be 0 using symmetry considerations. We examine the symmetry table given for the previous problem to see which symmetry gives a non-zero integral for
$$\int \psi_f A_1 B_1 \, d\tau = \int \psi_f B_1 \, d\tau$$
which implies that ψ_f must have B_1 symmetry and response **(C)** is correct.

S-5. To which symmetry group does cyclopropane (C_3H_6) belong?

(A) C_{3v} (B) D_{3d} (C) D_{3h} (D) T_d

Knowledge Required: The structure of cyclopropane; the basic symmetry operations and how to use them to determine the symmetry group of a molecule.

Thinking it Through: Cyclopropane has a C_3 axis through the center of the molecule. There is a plane of symmetry through the molecular plane, therefore the molecule has D_{3h} symmetry. The correct response is **(C)**.

S-6. The $^3P_2$ state of the oxygen atom has the $1s^2 2s^2 2p^4$ configuration. To which excited state is an electronic transition allowed in the electric dipole approximation?

(A) $1s^2 2s^2 2p^4$, $^3P_1$ (B) $1s^2 2s^2 2p^4$, $^1D_2$

(C) $1s^2 2s^2 2p^4$, $^1S_0$ (D) $1s^2 2s^2 2p^3 3p^1$, $^3P_1$

Knowledge Required: How to translate a term symbol into the correct L and S values; the electronic dipole transition selection rules.

Thinking it Through: ΔS must be 0, so the transition corresponding to responses **(B)** or **(C)** is not allowed. ΔL may be 0, so either transition **(A)** or transition **(D)** might be allowed. The transition in **(A)** has $\Delta l = 0$, and is forbidden. Response **(D)** is correct.

S-7. The planar methyl free radical, CH_3, is in the D_{3h} symmetry group. In this symmetry group, how many planes of symmetry are there?

(A) 3   (B) 4   (C) 6   (D) 7

Knowledge Required: Structure of molecules, visualization of molecular shapes from structure, determination of symmetry elements for molecules.

Thinking It Through: There is actually an abundance of information given in this problem. The molecule is shown with an electron in a p orbital perpendicular to the plane of the molecule. That is what puts the molecule in the D_{3h} symmetry group. There are three σ_v symmetry planes perpendicular to the plane of the molecule containing each of the C–H bonds. In addition, there is a σ_h plane (the plane of the molecule). Thus, there are four planes of symmetry, and the correct response is response **(B)**.

S-8. The combinations of atomic orbitals used in constructing appropriate molecular orbitals for BeH_2

(A) must have the same principal quantum number n.

(B) must have the same symmetry as the molecule.

(C) must have the same angular momentum quantum number, l.

(D) must be orthogonal.

Symmetry

Knowledge Required: Construction of symmetry adapted molecular orbitals, molecular orbitals as combinations of atomic orbitals, molecular orbital formation from atomic orbitals if similar energy.

Thinking It Through: The principal quantum number for the valence (bond forming) electrons of beryllium is 2; while that of hydrogen is 1; thus response **(A)** is incorrect. While it is true that the lowest occupied orbital for beryllium is $2s$, the bonding orbital would have a different angular momentum than the s orbital of the hydrogen atom; thus response **(C)** is incorrect. If the atomic orbitals were orthogonal; then there would be no overlap and no bond formed; thus response **(D)** is incorrect. The orbitals must be of the same symmetry, thus response **(B)** is the correct response.

S-9. *Trans*-1,2-dibromoethene possesses which elements of symmetry?

(A) C_2, σ_v (B) C_2, σ_v, i (C) C_2, σ_v, σ_h (D) C_2, σ_h, i

Knowledge Required: Shape of trans-1,2-dibromoethane, symmetry elements relative to molecular shape.

Thinking It Through: Trans-1,2-dibromoethane is a planar molecule, and it contains a center of symmetry (i) at the middle of the C=C bond. Thus responses **(A)** and **(C)** are incorrect. The C_2 axis is indicated on the figure (which is perpendicular to the page and runs through the middle of the C=C bond). A σ_v plane would contain the C_2 axis, while a σ_h plane would be perpendicular to the C_2 axis. For this molecule a σ_h plane (the plane of the molecule) is present, thus response **(D)** is the correct response and response **(A)** is incorrect.

S-10. In the molecule allene, the C=C=C group is linear and the terminal CH$_2$ groups are at right angles to each other, i.e. H$_1$ and H$_2$ lie in the yz plane and H$_3$ and H$_4$ in the xz plane:

The double bond between C$_1$ and C$_2$ is best described as involving the set of atomic and hybrid orbitals:

(A) $C_1(sp^2), C_2(sp), C_1(p_x), C_2(p_x)$ (B) $C_1(sp^2), C_2(sp), C_1(p_y), C_2(p_y)$

(C) $C_1(sp^2), C_2(sp^2), C_1(p_x), C_2(p_y)$ (D) $C_1(sp^2), C_2(sp), C_1(p_x), C_2(p_y)$

Knowledge Required: Formation of molecular orbitals from atomic orbitals.

Thinking It Through: A double bond is formed from an sp^2 hybrid orbital on C$_1$ and an sp hybrid orbital on C$_2$ in the bond and a p orbital perpendicular to the hybrid orbital from each atom. Thus we look for the response that has C$_1$ sp^2 orbital and C$_2$ sp orbitals. Only response **(A)** meets that criteria, so it must be the correct response.

Practice Questions

1. The water molecule belongs to the C_{2v} symmetry group which contains **only**

    (A) a 90 and 180 degrees rotation axis.
    (B) a 180 degrees rotation axis, two reflection planes, and the identity operation.
    (C) a 180 degrees rotation axis and the identity operation.
    (D) a reflection plane and the identity operation.

2. A molecule of C_{2v} symmetry is in an initial state characterized by A$_2$ symmetry. What is the symmetry of the final state when an IR transition in the x-direction is allowed? The character table for C_{2v} point group is given in study question S-3.

    (A) A$_1$ (B) A$_2$ (C) B$_1$ (D) B$_2$

Symmetry

3. Wave functions and electronic states of heteronuclear diatomic molecules are not labeled "ungerade" or "gerade" because these molecules

   (A) have no center of inversion.

   (B) have a permanent dipole moment.

   (C) have only two independent axes of rotation.

   (D) have no symmetry with respect to a plane containing the internuclear axis.

4. The molecule in the same symmetry group as NH_3 is

   (A) BF_3 (planar)   (B) CH_4

   (C) CH_3OH (D) CCl_3Br

5. In this configuration, ferrocene does *not* contain which symmetry element?

   (A) σ_v (B) C_5 (C) i (D) σ_h

6. Which molecule has energy levels described by an asymmetric top?

   (A) CH_4 (B) $HCCl_3$

   (C) C_2H_2 (D) BF_3

7. How many normal modes are present for H_2O_2?

   (A) 4   (B) 6   (C) 7   (D) 12

8. Molecules of the point group D_{2h} cannot be chiral. Which symmetry element rules out chiral molecules in this point group?

   (A) i (B) C_2 (C) σ_h (D) σ_v

9. Molecules in which point group exhibit a C_4 axis?

   (A) T_d (B) O_h (C) $D_{\infty h}$

   (D) none of these exhibit a C_4 axis.

10. The number of nondegenerate irreducible representations for the C_{2v} point group is

    (A) 0   (B) 1   (C) 2   (D) 4

Answers to Study Questions

1. A
2. D
3. D
4. C
5. C
6. D
7. B
8. B
9. D
10. A

Answers to Practice Questions

1. B
2. D
3. A
4. D
5. C
6. B
7. B
8. C
9. B
10. D

Molecular Orbital Theory

The solutions to the Schrödinger equation for the hydrogen-like atom are important to chemists because they are the building blocks, both conceptually and mathematically, for wave functions of molecules. As with the one-electron atoms, we begin with the Schrödinger equation, but because it cannot be solved exactly for multi-electron systems, we must make some approximations. The full Hamiltonian for a molecule has the form

$$\hat{H} = \hat{T}_{electrons} + \hat{T}_{nuclei} + \hat{V}_{nuclei/electrons} + \hat{V}_{electron/electron} + \hat{V}_{nuclei/nuclei}$$

The electron-electron repulsion term, $\hat{V}_{electron/electron}$, is what makes the problem impossible to solve analytically. The Born-Oppenheimer approximation, where the kinetic energy of the nuclei is ignored in finding the wave functions of the electrons, is frequently used by chemists to simplify molecular problems. Under this approximation the Hamiltonian for the electrons becomes

$$\hat{H}_{electronic} = \hat{T}_{electrons} + \hat{V}_{nuclei/electrons} + \hat{V}_{electron/electron} + \hat{V}_{nuclei/nuclei}$$

where the potential term involving only the nuclei is a constant.

There are conceptual advantages to considering approximate solutions found using variational theory. A trial wave function can be constructed by taking a linear combination of the ground state hydrogen atomic wave functions

$$\psi_{H_2^+} \approx N\left(\psi_{1sA} + \psi_{1sB}\right)$$

where N is a normalization constant and A and B denote the two hydrogen nuclei. This function exhibits a larger electron probability between the nuclei than on the perimeter of the molecule and is therefore considered "bonding". This molecular orbital (MO) is often designated as $1s\sigma_g$, where σ indicates the electron density is symmetric around the bond axis, the subscript g indicates the MO is *gerade* or even with respect to the center of symmetry in the molecule. The corresponding LCAO (linear combination of atomic orbitals) wave function for the first excited state of H_2^+ is

$$N\left(\psi_{1sA} - \psi_{1sB}\right)$$

typically designated $1s\sigma_u^*$. This MO has a node in the electron density between the atoms, hence it is characterized as *ungerade* (odd) and is antibonding (indicated by the *).

Approximate wave functions for higher energy states of H_2^+ can be constructed similarly. We can generate a molecular orbital energy diagram from the results. Just as a (slightly modified) hydrogen atom orbital energy diagram is used to predict the electron configurations and properties of multi-electron atoms, this MO energy diagram can be used for multi-electron diatomic molecules. A molecular wave function can be constructed from the orbitals, but it is a non-trivial process and generally requires the use of a computer. Despite this, molecular orbital diagrams can be used to correctly predict the triplet nature of O₂ and the instability of He₂. Bond orders can easily be estimated using

$$\text{bond order} = \frac{\text{\# of bonding e}^- - \text{\# of anti-bonding e}^-}{2}$$

Useful as the LCAO-MO conceptual framework described can be, it does not give the full functional form of the molecular wave function from which we could extract the total electronic energy or other properties, such as the dipole moment. There are many approaches which can be used to find a complete molecular wave function. One which arises naturally from the framework just developed uses the Slater determinant to construct a wave function from the MOs that is antisymmetric with respect to interchange of the electrons. For example, consider a Slater determinant wave function for He₂. Each entry in the determinant is a product of a spin function (α or β) and a spatial LCAO-MO. The number in parentheses designates the electron the function is describing.

Molecular Orbital Theory

$$\psi_{He_2} \approx \begin{vmatrix} \sigma_g(1)\alpha & \sigma_g(1)\beta & \sigma_u^*(1)\alpha & \sigma_u^*(1)\beta \\ \sigma_g(2)\alpha & \sigma_g(2)\beta & \sigma_u^*(2)\alpha & \sigma_u^*(2)\beta \\ \sigma_g(3)\alpha & \sigma_g(3)\beta & \sigma_u^*(3)\alpha & \sigma_u^*(3)\beta \\ \sigma_g(4)\alpha & \sigma_g(4)\beta & \sigma_u^*(4)\alpha & \sigma_u^*(4)\beta \end{vmatrix}$$

The variational theorem can be used to find the best approximation for this form of wave function by finding the set of LCAO coefficients (the c_1, c_2 etc.) that give the lowest energy. Clearly this is a computationally demanding task even for small molecules!

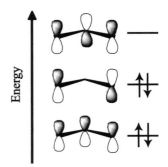

LCAO-MOs can be constructed for polyatomic molecules as well. MOs describing the π orbitals of a conjugated polyene can be built from the appropriate set of p orbitals. Consider the π orbitals of allyl anion $\left([H_2CCHCH_2]^- \right)$ shown at the right. Note that the lowest energy linear combination has only one node (in the plane of the molecule). It is generally true that as the number of nodes increases, so does the energy.

Study Questions

MO-1. For the H_2 molecule, which spatial molecular orbital has both covalent and ionic terms? Note that $\psi_{1sA}(1)$ represents electron 1 in the $1s$ orbital of hydrogen atom A.

(A) $c_1\psi_{1sA}(1)\psi_{1sB}(2)$

(B) $c_1\psi_{1sA}(1)\psi_{1sB}(2) + c_2\psi_{1sA}(2)\psi_{1sB}(1)$

(C) $c_1\psi_{1sA}(1)\psi_{1sB}(1) + c_2\psi_{1sA}(2)\psi_{1sB}(2)$

(D) $c_1\psi_{1sA}(1)\psi_{1sB}(2) + c_2\psi_{1sA}(2)\psi_{1sB}(1) + c_3\psi_{1sA}(1)\psi_{1sA}(2) + c_4\psi_{1sB}(1)\psi_{1sB}(2)$

Knowledge Required: Ionic contributions will be distributed asymmetrically over a molecule, by having multiple electrons on a single atom.

Thinking it Through: Each of the terms in response **(B)** has an electron on both hydrogen atom A and B, so **(B)** has only covalent character. The terms in **(C)** are likewise distributed over both centers. The first and second terms in **(D)** are covalent (as in response **(B)**), but the third and fourth terms each have only contributions from one center (the third term is on center A and the fourth on center B). These terms are ionic contributions, therefore the correct response is **(D)**. Response **(A)** only contains a single term and is not a valid wave function.

MO-2. In which diatomic molecule are all the electron spins paired?

(A) CO (B) NO (C) O_2 (D) FO

Molecular Orbital Theory

Knowledge Required: The principles followed in filling an electron diagram: the Aufbau principle (fill from the bottom up); Hund's rule (fill across a degenerate level before beginning to pair electron spins); Pauli exclusion principle (only 2 electrons in an orbital, of different spins).

Thinking it Through: NO and FO have an odd number of electrons, so cannot possibly have all their spins paired. Therefore neither of responses **(B)** nor **(D)** is correct. The electron configuration of CO is $(1s\sigma_g)^2(1s\sigma_u^*)^2(2s\sigma_g)^2(2s\sigma_u^*)^2(2p\pi_u)^4(2p\sigma_g)^2$ which has all the electrons paired, so the correct response is response **(A)**. The electron configuration of O_2 is $(1s\sigma_g)^2(1s\sigma_u^*)^2(2s\sigma_g)^2(2s\sigma_u^*)^2(2p\pi_u)^4(2p\sigma_g)^2(2p\pi_g^*)^2$ where the highest occupied level is degenerate with one electron in each orbital, and unpaired spins. You might also have recalled the common piece of chemistry trivia that molecular oxygen is paramagnetic and therefore has unpaired spins!

MO-3. The diatomic molecule formed from Li and Li^+ is predicted to be stable from molecular orbital theory because it has

(A) three electrons in bonding orbitals and two in antibonding orbitals.

(B) three electrons in bonding orbitals and three in antibonding orbitals.

(C) three electrons in bonding orbitals and one in antibonding orbitals.

(D) all electrons in bonding orbitals.

Knowledge Required: The principles followed in filling an electron diagram; stable molecules have a bond order that is greater than zero.

Thinking it through: The Li_2^+ molecule has 5 electrons, so we can immediately reject responses **(B)** and **(C)**. Examination of the MO energy diagram for diatomics reveals that Li_2^+ has a configuration of $(1s\sigma_g)^2(1s\sigma_u^*)^2(2s\sigma_g)^1$, with 3 electrons in bonding orbitals and 2 in antibonding orbitals. Therefore the correct response is response **(A)**.

MO-4. Using the simple homonuclear molecule MO energy level scheme for the ground state of the diatomic boron molecule, we would predict that

(A) B_2^{2+} will have unpaired electrons. (B) B_2 will have all electrons paired.

(C) B_2 will have unpaired electrons. (D) B_2^{2-} will have unpaired electrons.

Knowledge Required: The principles followed in filling an electron diagram.

Thinking it Through: B_2^{2+} has the configuration $(1s\sigma_g)^2(1s\sigma_u^*)^2(2s\sigma_g)^2(2s\sigma_u^*)^2$, therefore all the electrons are paired and response **(A)** is incorrect. B_2 has the configuration $(1s\sigma_g)^2(1s\sigma_u^*)^2(2s\sigma_g)^2(2s\sigma_u^*)^2(2p\pi_u)^2$ where the electrons in the $2p\pi_u$ level are unpaired. Therefore response **(C)** is correct and response **(B)** is incorrect. Adding two electrons to the neutral molecule will fill the $2p\pi_u$ level and all electrons will be paired, so response **(D)** is incorrect.

MO-5. When an electron is removed from N_2, the N–N bond lengthens; whereas when an electron is removed from O_2, the O–O bond shortens. The most important factor involved in this difference between nitrogen and oxygen is that

(A) oxygen is more electronegative than nitrogen.

(B) the bond distance in neutral nitrogen is shorter than in oxygen.

(C) the electron comes from an antibonding MO in oxygen but from a bonding MO in nitrogen.

(D) oxygen has a higher ionization potential than nitrogen.

Knowledge Required: The principle followed in filling an electron diagram; bond order calculations.

Thinking it Through: Referring to the homonuclear diatomic molecular orbital diagram given in the introduction, we see that N_2 has a bond order of 3 and an electron would be removed from the $2p\sigma_g$ orbital (a bonding orbital) while O_2 has a bond order of 2 and an electron would be removed from the $2p\pi_g^*$ orbital (an antibonding orbital). Removal of an electron from N_2 will decrease the bond order; lengthening the bond, while removal from O_2 will increase the bond order; shortening the bond. Thus response **(C)** is the most correct response. While responses **(B)** and **(D)** are correct observations, none of responses **(A)**, **(B)**, or **(D)** provides an explanation.

MO-6. The diagrams shown each represent the atomic orbitals that describe the π electron system in butadiene; shaded areas are positive – unshaded areas are negative. Which of these diagrams would be expected to represent the *highest* energy molecular orbital?

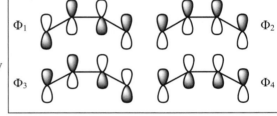

(A) Φ_1 (B) Φ_2 (C) Φ_3 (D) Φ_4

Knowledge Required: How the number of nodes changes with increasing energy.

Thinking it Through: For atomic and molecular systems, the energy increases as the number of nodes increases. This effect is seen in atomic systems with increasing n and l. Here we look at molecular systems. All of these molecular orbitals have a node in the plane of the molecule, so we are left to count nodes perpendicular to the molecular plane. The wave function Φ_1 has 3 nodes perpendicular to the plane; Φ_2 has zero; Φ_3 has 1; and Φ_4 has 2. The energy ordering of these wave functions is then $\Phi_2 < \Phi_3 < \Phi_4 < \Phi_1$; so that Φ_1 is the highest energy and response **(A)** is the correct response.

MO-7. Which expression is the best choice for the spatial part of a molecular orbital wave function for the ground state for H_2? (ψ for the 1s atomic orbital of atom **A** containing electron one is written as $1s_A(1)$.)

(A) $1s_A(1)1s_B(2) + 1s_B(1)1s_A(2)$

(B) $1s_A(1)1s_B(2) - 1s_B(1)1s_A(2)$

(C) $1s_A(1)1s_B(2)1s_B(1)1s_A(2)$

(D) $[1s_A(1) + 1s_B(1)][1s_A(2) + 1s_B(2)]$

Knowledge Required: Form of the LCAO wave function for H_2.

Thinking it Through: The ground state wave function has an asymmetric spin part; which eliminates response **(B)** as incorrect. Response **(C)** is also incorrect by having the incorrect form with multiplication of two wave functions for each electron. Response **(D)** includes both covalent and ionic terms, so it is incorrect. Thus the covalent only expression given by response **(A)** is correct.

Molecular Orbital Theory

MO-8. Because the nuclear motions are much slower than those of the electron, the molecular Schrödinger equation for the electron motion can be solved by assuming that the nuclei are at fixed locations. This is

(A) the Born-Oppenheimer approximation. (B) the time-dependent Schrödinger equation.
(C) Russell-Saunders coupling. (D) the variation method.

Knowledge Required: The meanings and application of the various terms.

Thinking it Through: The Born-Oppenheimer approximation separates nuclear and electronic motion; and thus the correct response is response **(A)**. Response **(B)** is a correct theory name but describes the time dependence of the wave function. Response **(C)** describes the coupling between orbital and spin angular momentum and is thus incorrect. Response **(D)** is related to performing calculations and is thus incorrect.

MO-9. For the π electrons in a polymethine dye having this formula, the simple Hückel molecular orbital method has the secular determinant equal to

(A) $\begin{vmatrix} x & 1 & 0 & 0 & 0 \\ 1 & x & 1 & 0 & 0 \\ 0 & 1 & x & 1 & 0 \\ 0 & 0 & 1 & x & 1 \\ 0 & 0 & 0 & 1 & x \end{vmatrix} = 0$

(B) $\begin{vmatrix} x & 1 & 0 & 0 & 1 \\ 1 & x & 1 & 0 & 0 \\ 0 & 1 & x & 1 & 0 \\ 0 & 0 & 1 & x & 1 \\ 1 & 0 & 0 & 1 & x \end{vmatrix} = 0$

(C) $\begin{vmatrix} x & 1 & 0 & 0 & 0 \\ 1 & x & 1 & 0 & 0 \\ 0 & 1 & x & 1 & 0 \\ 0 & 0 & 1 & x & 1 \\ 0 & 0 & 0 & 1 & x \end{vmatrix} = 1$

(D) $\begin{vmatrix} x & 1 & 1 & 1 & 1 \\ 1 & x & 1 & 1 & 1 \\ 1 & 1 & x & 1 & 1 \\ 1 & 1 & 1 & x & 1 \\ 1 & 1 & 1 & 1 & x \end{vmatrix} = 0$

Knowledge Required: The form of the Hückel wave functions.

Thinking it Through: A Hückel wave function has x along the diagonal elements connecting to neighbor atoms are 1; with all the other being 0. Thus the correct form corresponds to response **(A)**.

MO-10. A molecular orbital of ethylene is shown. This orbital is

(A) π_g. (B) π_u. (C) σ_g. (D) σ_u.

Knowledge Required: Orbital designations within bonds and their shapes.

Thinking it Through: The orbital is a π orbital because it extends above and below the plane of the molecule. There is a center of symmetry making this a gerade (g) orbital. Thus the correct response is response **(A)**.

Practice Questions

1. According to molecular orbital theory, which molecule should be least stable?

   (A) Li_2 (B) Be_2 (C) H_2 (D) C_2

2. The electronic ground state of the O_2^- molecular ion may be described the molecular orbital configuration $(1s\sigma_g)^2 (1s\sigma_u^*)^2 (2s\sigma_g)^2 (2s\sigma_u^*)^2 (2p\sigma_g)^2 (2p\pi_u)^4 (2p\pi_g^*)^3$ where the superscript denotes an antibonding orbital. The bond order for O_2^- would therefore be

   (A) 1 (B) 2 (C) 3/2 (D) 5/2

3. The diagrams represent the π molecular orbitals of *trans*-butadiene formed from linear combinations of the carbon *p* orbitals. The algebraic sign of the *p* orbital is indicated by shading. Which molecular orbital is the lowest unoccupied molecular orbital (LUMO) of *trans*-butadiene which has 4 π electrons?

   (A)  (B)

   (C)  (D)

4. The highest occupied molecular orbital for benzene with 6 π electrons is best described by which figure?

   (A)  (B)

   (C) 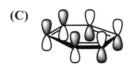 (D)

5. Which molecule has a bond order which is different from the others?

   (A) NO (B) O_2^+
   (C) N_2^{2+} (D) CO^-

6. The Cr_2 molecule has a bond length of 0.16 nm. What is the bond order?

   (A) 0 (B) 1 (C) 3 (D) 6

7. The ground state of the NO molecule is a $^2\Pi$ state. What is the electron configuration for this state?

   (A) $...3\sigma^2\ 1\pi^4\ 1\pi^{*1}\ 3\sigma^{*0}$
   (B) $...3\sigma^2\ 1\pi^4\ 1\pi^{*0}\ 3\sigma^{*1}$
   (C) $...3\sigma^2\ 1\pi^3\ 1\pi^{*2}\ 3\sigma^{*0}$
   (D) $...3\sigma^2\ 1\pi^3\ 1\pi^{*0}\ 3\sigma^{*2}$

8. Distinguishing features of a σ^* orbital compared to a σ orbital include

   (A) both are delocalized.
   (B) both have nodal planes perpendicular to the bond axis.
   (C) both have cylindrical symmetry about the bond axis.
   (D) the antibonding orbital is asymmetric about the bond axis.

9. You have just obtained the photoelectron spectrum of O_2. When compared with the vibrational frequency of the O_2 precursor, the vibrational frequency of the resulting O_2^+ is

   (A) greater. (B) less.
   (C) the same. (D) unpredictable.

10. Which molecule will have the highest **first** ionization energy?

    (A) C_2 (B) N_2 (C) NO (D) O_2

Molecular Orbital Theory

Answers to Study Questions

1. D
2. A
3. A
4. C
5. C
6. A
7. A
8. A
9. A
10. A

Answers to Practice Questions

1. B
2. C
3. D
4. B
5. C
6. D
7. A
8. C
9. A
10. B

Spectral Properties

Understanding spectra is an important reason for studying quantum chemistry. Typically you learn about models describing nuclear spin, rotation, vibration, and electronic properties of molecules and the electronic properties of atoms. Each model is associated with a region of the electromagnetic spectrum where the lines associated with transitions between quantum states are observed. Here we summarize some of the concepts underpinning observed spectra.

Microwave spectra are due to transitions between rotational quantum states. Typical spectra of diatomic molecules show a series of equally spaced lines with varying intensity. The distance between energy levels of rotor quantum states increases with increasing J quantum number. Transitions between these states give rise to the observed lines. The factors controlling the intensity of lines in the spectrum include the Boltzmann distribution across the closely spaced rotational states and the degeneracy of the angular momentum states.

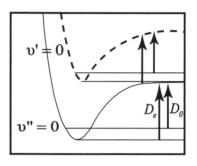

Vibrations for diatomic molecules are detected in the IR region. A force constant directly proportional to the bond tightness characterizes all bonds. Rotational transitions accompany vibrational transitions. They are seen as fine structure under higher resolution. One can also see additional fine structure due to the presence of atomic isotopes or the presence of a magnetic field. Selection rules tell us that a vibrational transition can only occur when the change in quantum number between two states, $\Delta \upsilon$ is ± 1. Simultaneously the change in quantum number for rotational states associated with the two vibrational states is restricted to $\Delta J = \pm 1$. This produces two branches in the typical IR spectrum: an R branch with $\Delta J = +1$ and a P branch with $\Delta J = -1$. The lines in the spectrum are nearly equally spaced on either side of the Q branch. There is no Q branch unless $\Delta J = 0$. A non-zero transition moment must exist for a line to appear in a spectrum.

Molecules also show spectra that appear in the visible or ultraviolet region arising from transitions between electronic quantum states. Each electronic transition is accompanied by vibrational and rotational transitions. Each has its own selection rules determined by a non-zero transition moment that is governed by the symmetry of the electronic state. Each electronic state has its own potential energy curve characterized by ν_0, D_e, and D_0. One can see a progression corresponding to transitions from $\upsilon'' = 0$ to $\upsilon' = 0, 1, 2, 3\ldots$ or a series of lines with $\Delta \upsilon = 0$, i.e. $\upsilon'' = 0 \rightarrow \upsilon' = 0, \upsilon'' = 1 \rightarrow \upsilon' = 1 \ldots$ depending on the change in internuclear distances of the two electronic states. The intensity of the lines is determined by Franck-Condon factors which are proportional to the product, or overlap, of the wave functions in the ground and excited states. There are selection rules that govern electronic transitions of atoms, $\Delta L = \pm 1$, $\Delta S = 0$, and $\Delta J = 0, \pm 1$, where L is the orbital angular momentum quantum number, S is the total spin quantum number, and J is the total angular momentum number. Allowed transitions include S $\leftrightarrow$ P, P $\leftrightarrow$ D, and D $\leftrightarrow$ F. Forbidden transitions include S $\leftrightarrow$ S ($\Delta L = 0$) and S $\leftrightarrow$ D ($\Delta L = +2$). As atomic numbers increase, the selection rules break down and atomic spectra become more complex as weak lines appear that are due to forbidden transitions.

Raman spectra consist of sets of lines due to an exchange of energy between a molecule and the incident exciting frequency during the scattering of the incident beam. Most of the incident light is scattered at the original frequency. Some is scattered at higher frequencies (anti-Stokes lines) and some is scattered at lower frequencies (Stokes lines). Stokes lines are typically more intense than anti-Stokes lines. Rotational Raman lines lie close to the exciting frequency while vibrational Raman lines lie further from the exciting frequency because of the larger energy gap between the two vibrational states. For a molecule to display Raman spectral lines, the mode of vibration must be polarizable by the electric field of the incident radiation. Molecules possessing a center of symmetry will have vibrational frequencies that are either Raman active or IR active but not both.

Spectral Properties

Study Questions

SP-1. The intensity of an absorption line is typically determined by

(A) the wavelength of the transition.

(B) the value of the transition moment integral.

(C) only the lifetime of the final state.

(D) only the population of the final state.

Knowledge Required: Factors governing the appearance of a spectral line and determining the line intensity for various spectral types. Appearance of a line requires that the transition moment be non-zero. The selection rules for various types of transitions between quantum states. For transitions between electronic quantum states the Franck-Condon Principle, that the transition integral be non-zero, applies and determines the intensity of a peak. The transition integral depends on the overlap between the vibrational wave functions of the excited and the ground states. The greater the overlap the more intense is the observed peak.

Thinking it Through: Although the wavelength of the incident light can be matched precisely to the energy difference between two quantum states, there will be no transition and no spectral peak unless the transition moment is non-zero. Thus response **(A)** is incorrect. Response **(C)** would be correct for fluorescence but even then the transition moment must be non-zero. We might also consider response **(D)** because a state must be populated in order for it to interact with an electric field and show a transition to a higher state. In any case only the population of the initial state controls the probability of a transition, the population of the final state is immaterial. However, even a very populated state will not give a spectral peak if the transition moment is zero. Response **(D)** is incorrect. The correct answer is response **(B)**.

SP-2. Transitions that represent only changes in rotational energy appear in which spectral range?

(A) Microwave (B) Infrared (C) Visible (D) X-ray

Knowledge Required: That each spectral region represents an energy range and that the relative magnitudes for spacing between quantum states are rotational < vibrational < electronic respectively corresponding to the microwave, IR, and UV-vis regions of the electromagnetic spectrum respectively.

Thinking it Through: Response **(D)** is incorrect because X-rays are very energetic and can ionize atoms and eject electrons from the inner shell of an atom or molecule. Response **(C)** is incorrect because this is the energy range for exciting a molecule from one electronic state to another. Response **(B)** is incorrect because this is the frequency range that matches the ΔE between vibrational quantum states. Response **(A)** is correct. This frequency range matches the energy level separation between the rotational states of a molecule.

SP-3. A Raman spectrum of liquid carbon tetrachloride was obtained using the 331.0 nm line of a He/Cd laser as the excitation source. The maximum of the first band on the long wavelength side of this line was detected at 339.7 nm. What accounts for its presence?

(A) It must be a side band, since spherical top molecules do not possess a dipole moment.

(B) It arises from a rotational transition.

(C) It arises from a vibrational transition.

(D) It arises from an electronic transition.

Knowledge Required: Know roughly the wavelengths of the various spectral regions, e.g. IR is from 200 to 4000 cm^{-1}. To be able to calculate the frequency of the transition produced by the incident beam. That the observed line is due to a transition from an energy level given by the difference between the frequency of the incident light and the frequency of the observed spectral peak.

Thinking it Through: First compute the ΔE between the incident beam and the observed spectral line. This is given by $\tilde{\nu} = \dfrac{1}{331.0 \times 10^{-9} \text{ m}} - \dfrac{1}{339.7 \times 10^{-9} \text{ m}} = 773 \text{ cm}^{-1}$. The frequency computed for this line is in the IR region. Thus response **(A)** is not correct. Response **(B)** is not correct because the molecule has no dipole moment and thus there would be no rotational transitions allowed. Response **(D)** is incorrect because the observed line falls in the IR region and and electronic transitions generally fall in the UV-vis region. Response **(C)** is correct because the frequency of the observed line corresponds to a ΔE_ν for the energy difference between two vibrational states.

SP-4.

The figure represents a medium resolution UV–vis absorption spectrum of I_2(g) at 70 °C. The structure of the absorption band is due to transitions from

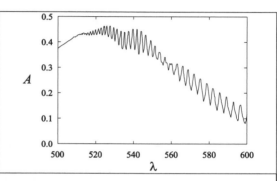

(A) the lowest vibrational level of the ground electronic state to several vibrational levels of several excited electronic states.

(B) the lowest vibrational level of the ground electronic state to several vibrational levels of a single excited electronic state.

(C) several vibrational levels of the ground electronic state to several vibrational levels of a single excited electronic state.

(D) several vibrational levels of the ground electronic state to the lowest vibrational level of a single excited electronic state.

Knowledge Required: The visible region corresponds to transitions between electronic states of a molecule. In a medium resolution UV-vis absorption spectrum one sees fine structure arising from vertical transitions from the $\upsilon" = 0, 1, 2$ vibrational states of the ground electronic state to higher, $\upsilon' = \ldots 25, 26, 27 \ldots$ vibrational states of the excited electronic state.

Thinking it Through: The figure shows a series of spectral lines characteristic of a UV-vis spectrum for I_2 from multiple vibrational levels of the ground state to multiple vibrational levels in the excited state. We can eliminate response **(A)** because it says that all peaks are due to transitions that originate at $\upsilon" = 0$. Response **(B)** also discusses several different excited states and this too is incorrect. Likewise response **(D)** is incorrect. The series of lines in a progression are due to excitation to the several vibrational states of the excited state. Response **(C)** is the correct answer.

| SP-5. | For a diatomic molecule a part of the potential energy diagram is shown with the approximate ψ^2 curves superimposed on the lines indicating the energy values for states. Which of these transitions would have the highest probability according to the Franck-Condon principle? | 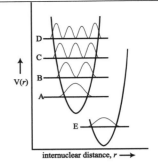 |

(A)   E → A   (B)   E → B   (C)   E → C   (D)   E → D

Knowledge Required: That the two figures represent the potential energy curves for two different electronic states of the molecule. The figures show the vibrational levels for each electronic state and the probability density for each level. The Franck-Condon Principle states that the most intense line would be given by the greatest overlap between the ground state $\upsilon = 0$ function and one of the excited state functions.

Thinking it Through: The figure shows the potential energy curves for two different electronic states of a molecule. The ground state appears to the right and below the excited state. Only the $\upsilon'' = 0$ wave function of the ground state, labeled E, is shown since this would be the most populated vibrational state at room temperature. The Franck-Condon Principle says that the transitions to excited state vibrational levels are vertical because the atoms in the molecule do not move during a transition. Consequently the most intense line would be given by the greatest overlap between the ground state $\upsilon'' = 0$ function and one of the excited state's probability densities. Examine the figure and draw a vertical transition line up from the maximum in the ground state function. Although the image is difficult to use the greatest overlap would occur between the ground state E and the excited state at vibrational level D. The correct response is thus **(D)**.

| SP-6. | What is the correct ordering for typical gaps between adjacent energy levels? |

     (A) $\Delta E_{electronic} > \Delta E_{vibrational} > \Delta E_{rotational} > \Delta E_{nuclear\ spin}$

     (B) $\Delta E_{electronic} > \Delta E_{nuclear\ spin} > \Delta E_{vibrational} > \Delta E_{rotational}$

     (C) $\Delta E_{nuclear\ spin} > \Delta E_{electronic} > \Delta E_{rotational} > \Delta E_{vibrational}$

     (D) $\Delta E_{nuclear\ spin} > \Delta E_{rotational} > \Delta E_{electroinc} > \Delta E_{vibrational}$

Knowledge Required: The size of the energy level separations for each type of quantum state transition. It doesn't take much energy to change the nuclear spin. Rotational energy levels are more widely spaced than nuclear spin states. The next largest are vibrational spacings and the largest are electronic spacings.

Thinking it Through: First write down the order of energy level spacings given the information above. Starting with the largest the spacings in decreasing order are electronic > vibrational > rotational > nuclear spin. The response **(A)** is immediately seen to be correct. All other responses have an error in the order.

| SP-7. | Which transition is allowed in the normal electronic emission spectrum of an atom? |

     (A) $2s \rightarrow 1s$ (B) $2s \rightarrow 3p$ (C) $3d \rightarrow 2p$ (D) $5d \rightarrow 2s$

Spectral Properties

Knowledge Required: The selection rules for electronic transitions between energy levels of atoms.

Thinking it Through: The selection rules that govern electronic transitions of atoms are $\Delta L = \pm 1$, $\Delta S = 0$, and $\Delta J = 0, \pm 1$. In response **(A)**, the L for each state is 0; ΔL is 0 and this transition is not allowed. In response **(B)**, L is 0 and 1 for each state respectively; ΔL is $+1$ and the transition is allowed. However this is an absorption transition. Response **(B)** is not the correct answer. In response **(C)** we have a value of $\Delta L = -1$. This corresponds to emission and satisfies the selection rule. Response **(C)** is the correct answer. Response **(D)** is incorrect because although it corresponds to emission it does not satisfy the selection rule because $\Delta L = -2$.

SP-8. A transition between two states with wavefunctions ψ_1 and ψ_2 is said to be electric-dipole allowed with polarization in the x direction if the dipole transition integral

(A) is 1. (B) is 0. (C) is non-zero. (D) is real.

Knowledge Required: The intensity of a transition between two states is proportional to the square of the absolute value of the transition integral, $\left| \int \psi_2^* x \psi_1 \, dx \right|^2$.

Thinking it Through: Since the intensity is proportional to the square of the transition integral, then any non-zero value for the transition integral will result in an observed spectral peak. Response **(A)** is incorrect because it gives only one value for the integral. Response **(B)** is incorrect because if the transition integral were zero then no spectral peak would be observed. Response **(D)** is also incorrect because the real number set includes all positive and negative numbers including zero. An observed transition cannot have a zero transition integral. Since the only constraint on the transition integral for an observed transition must be that it is non-zero, response **(C)** is correct. The square of any non-zero number is a positive number and this fits the requirement for an observed spectral line.

SP-9. Which spectrum depends for its existence on the polarizability of a molecule?

(A) Microwave (B) Raman (C) Infrared (D) Ultraviolet

Knowledge Required: The general rules governing the appearance of spectra in the various parts of the electromagnetic spectrum.

Thinking it Through: Selection rules are determined by evaluation of transition moments which in turn depend on the dipole moment or polarizability of a bond in a diatomic molecule or mode of vibration in a polyatomic molecule. In order to observe microwave spectra a molecule must have a dipole moment. Thus response **(A)** is incorrect. In order for a line to appear in the IR, the mode of vibration must have a changing dipole moment. For example, CO_2 has three modes of vibration. Two of these have changing dipole moments during the vibration and thus CO_2 has two observable IR modes. Thus response **(C)** is also incorrect. Spectra appearing in the UV-vis region correspond to electronic transitions with different charge distributions for each state connected by the transition. For transitions in the UV-vis region the transition dipole depends on the overlap of wave functions in the ground state and the excited state (known as Franck-Condon factors), the redistribution of electrons in the excited state relative to the ground state, the parity and the symmetry of the ground and excited states. The correct response is **(B)**. Raman spectroscopy requires a changing polarizability.

SP-10. Which one of the following molecules does NOT absorb in the infrared?

(A) I_2 (B) CCl_4 (C) NH_3 (D) HI

Spectral Properties

Knowledge Required: Gross selection rules for infrared (vibrational) transitions.

Thinking It Through: The gross selection rule for vibrational transitions is that the molecule must have a changing dipole molecule during the transition. Both responses **(C)** and **(D)** are molecules with a dipole moment; thus they will absorb in the infrared. Response **(B)** (CCl_4) does not have a dipole moment, but will experience a dipole moment in certain asymmetric modes of vibration and will thus have an infrared transition, I_2 can not develop a dipole on transition; thus response **(A)** is correct.

Practice Questions

1. The most intense transition for the UV–vis spectrum of a diatomic molecule for which the two electronic potential energy curves are shown here is

   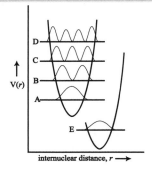

   (A) $v'' = 0 \to v' = 0$
   (B) $v'' = 0 \to v' = 2$
   (C) $v'' = 0 \to v' = 4$
   (D) $v'' = 0 \to v' = 6$

2. Which transition is forbidden under the selection rules for electric dipole radiation in the rigid rotor – harmonic oscillator approximation?

   (A) N_2 molecule, rotational transition from $J = 1$ to $J = 2$, $\Delta v = 0$
   (B) HF molecule, rotational transition from $J = 1$ to $J = 2$, $\Delta v = 0$
   (C) N_2 molecule, vibrational Raman transition from $v = 0$ to $v = 1$
   (D) N_2 molecule, electronic transition from the $^1\Sigma_g^+$ state to a $^1\Pi_u$ state

3. Which energy is the largest?

   (A) The typical energy of a single covalent bond.
   (B) The typical energy of a double bond.
   (C) The rotational barrier in C_2H_6.
   (D) The ionization energy of a hydrogen atom.

4. How many of the fundamental vibrational modes of the non-linear SO_2 molecule are IR active?

   (A) none (B) one
   (C) two (D) all

5. Given that the energy of an electron in a hydrogen atom is $-R_H/n^2$, the energy of the photon that can excite an electron from the ground state to the first excited state is

   (A) $(3/4)R_H$ (B) R_H
   (C) $(1/4)R_H$ (D) $(5/4)R_H$

6. In the harmonic oscillator approximation only a single line appears in the absorption spectrum for common molecules. Reason(s) for this include

   (A) the energy levels are equally spaced.
   (B) the selection rule is $\Delta v = +1$.
   (C) the molecule has a changing dipole moment.
   (D) all of the above.

7. The O atom has several emission features. Which will have the longest lifetime?

   (A) $2p^4\ ^1D \to 2p^4\ ^3P$
   (B) $2p^4\ ^1S \to 2p^4\ ^3P$
   (C) $2p^3 3p\ ^3P \to 2p^4\ ^3P$
   (D) $2p^3 3s\ ^3P \to 2p^4\ ^3P$

8. The O atom has several emission features. Which will have the lowest frequency?

(A) $2p^4\ ^1D \rightarrow 2p^4\ ^3P$

(B) $2p^4\ ^1S \rightarrow 2p^4\ ^3P$

(C) $2p^3 3p\ ^3P \rightarrow 2p^4\ ^3P$

(D) $2p^3 3s\ ^3P \rightarrow 2p^4\ ^3P$

9. The CH₃ radical is planar. Which mode is not infrared active?

(A) ... (B) ... (C) ... (D) ...

10. Which has the longest lifetime?

(A) electronic fluorescence

(B) electronic absorption

(C) intersystem crossing

(D) vibrational emission

Spectral Properties

Answers to Study Questions

1. B
2. A
3. C
4. C
5. D
6. A
7. C
8. C
9. B
10. A

Answers to Practice Questions

1. D
2. A
3. D
4. D
5. A
6. D
7. A
8. A
9. C
10. D

Advanced Topics: Electronic Structure Theory and Spectroscopy

While exact solutions may elude the chemist, useful approximate methods to find solutions have been developed. Perturbation theory and the variational theorem are two approaches that have found broad utility in chemistry. Perturbation theory treats the Hamiltonian as the sum of a Hamiltonian for which a solution is known and a small correction or perturbation to the known system. The energy and wave function of the perturbed system are written as a power series in terms of the unperturbed solutions, $\{\psi_i^{(0)}\}$. The series are typically truncated at some small n. The first order correction to the energy is

$$E_i^{(1)} = \int \psi_i^{(0)*} \hat{H}' \psi_i^{(0)} d\tau$$

so a perturbation approximation for the ith energy to first order would be

$$E_i \approx E_i^{(0)} + E_i^{(1)} = E_i^{(0)} + \int \psi_i^{(0)*} \hat{H}' \psi_i^{(0)} d\tau$$

Finding the perturbed first order wave function is somewhat more involved, since it often involves the infinite sum

$$\psi_i \approx \psi_i^{(0)} + \sum_{m \neq n} \frac{\int \psi_m^{(0)*} \hat{H}' \psi_n^{(0)} d\tau}{E_n^{(0)} - E_m^{(0)}} \psi_m^{(0)}$$

Perturbation theory has the advantage that it may be generally applied to excited states, and so is of particular use in spectroscopy.

Variational approaches begin by constructing a trial wave function that satisfies the boundary conditions of the problem. The full Hamiltonian is used and an estimate to the energy is given by

$$E_{varational} = \frac{\int \varphi_{trial}^* \hat{H} \varphi_{trial} d\tau}{\int \varphi_{trial}^* \varphi_{trial} d\tau} \geq E_{exact}$$

As long as an appropriate trial wave function satisfying the general requirements for wave functions, such as normalizability, and meeting the boundary conditions for the problem has been used, the variational estimate for the energy is guaranteed to be an upper bound to the exact energy. The energy can thus be used as a measure of the quality of the trial wave function, the lower the energy, the closer the solution is to the exact value. If the trial wave function includes an adjustable parameter, the best wave function of that form can be found by determining the value of the parameter that minimizes the variational energy. Linear variation theory creates a trial wave function from a linear combination of appropriate functions, e.g.

$$\varphi_{trial} = \sum_i c_i \varphi_i = c_1 \varphi_1 + c_2 \varphi_2 + \text{constant}$$

Generally, variation theory can only be applied to ground states.

Computational chemistry encompasses a wide variety of quantum mechanical electronic structure theories, as well as important methods based on classical mechanics. A passing familiarity with the basic principles behind each method as well as the associate acronyms is required for modern chemists.

MM (Molecular Mechanics). Though it draws on the harmonic oscillator model, MM uses classical mechanics for computing the energy of a molecule based on its molecular geometry. Searching for the geometry that gives the lowest energy yields a prediction of the molecular structure. Different force fields have been designed for different classes of molecules. A force field is defined by a set of force constants, atom types and the form of the equation used to estimate the energy. MM description depends on the bonds specified by the user and cannot describe bond- making or breaking processes. The quality of a structure and energy determined by MM depends on the quality of the force field used, and ultimately on the quality of the experimental information used to parameterize (choose the force constants) in the force field. MM methods are extremely fast.

MD (Molecular Dynamics). MD uses Newton's equations of motion to solve for the trajectory of a molecule or set of molecules (the movements of each atom) over time. The energy can be determined using molecular mechanics or quantum mechanics or a combination of both. MD is restricted to short time scales (typically picoseconds), which limits the types of processes that can be modeled.

Semi-empirical MO theory. Semi-empirical methods range from the sort that can be done by a persistent student on a napkin (Hückel MO theory) to those requiring large computing resources. The common thread in these methods is that they ignore some contributions to the electronic structure (such as non-valence electrons and overlap between certain atoms) and parameterize other contributions to experimental data. The more sophisticated methods within this class give very high quality results. The introduction of experimental data into the method means that it is technically not variational, but in practice, the energies are upper bounds to the true energy.

***ab initio* methods.** *Ab initio* methods do not rely on any experimental information other than the values of the fundamental constants such as Planck's constant. A word of warning, *ab initio* methods do not necessarily produce better results for a given system than HF and DFT are both *ab initio* methods.

HF MO theory. (Hartree-Fock MO theory) HF theories assume that the wave function has the form of a Slater determinant, where each orbital in the determinant is a linear combination of basis functions. The size and flexibility of the basis set controls the quality of the results. HF is a variational method, the more flexible the basis set, the better (lower) the energy. Lower energies do not mean, however, that other properties, such as molecular structure or dipole moment will be closer to experiment. Common basis sets range from the very small STO-3G basis to large and complex bases such as cc-pV6Z. HF wave functions are found using a self-consistent field (SCF) algorithm, which finds the best set of coefficients for each occupied molecular orbital in the average field of the other orbitals. While computationally effective, this method fails to account for all of the energy resulting from the correlated movements of the electrons. Post-HF methods have been developed which do a better job of computing this electron correlation energy, including CI (configuration interaction) and MP (Møller-Plesset) methods. The computational requirements for a HF calculation go up with the number of electrons and basis sets. Post-HF calculations are generally require greater computing time and space than HF alone.

DFT (density functional theory). Instead of formulating the problem in terms of a many-electron wave function, DFT is based on the total electron density function. Different functionals have been developed to describe the effective potential energy in which the electrons move, including B3LYP and Becke 88.

Lasers (light amplification by stimulated emission of radiation) are an example of practical quantum mechanics. The potential for lasing behavior was first predicted theoretically. Lasers can be made from many different materials. The schematic for a basic laser is shown below

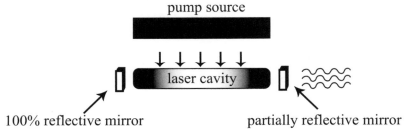

The lasing medium is confined in the cavity. The pump source excites the target molecule or atom in the cavity, creating a population inversion. As molecules return from the excited state, some of the emitted light is reflected back through the medium, stimulating further emission. The stimulated emissions are in phase with the light waves already reflecting inside the cavity. A coherent beam of light is produced, exiting through the partially reflecting mirror. Lasers are a standard tool for spectroscopists.

Electronic Structure Theory and Spectroscopy

Study Questions

EST-1. According to the variation principle, an improved approximate wave function is obtained by

(A) minimizing the electron density. (B) maximizing the electron density.
(C) minimizing the energy. (D) maximizing the energy.

Knowledge Required: The variational theorem; the variational energy is often used as a measure of the quality of the wave function.

Thinking it Through: The variational theorem states that the energy of any appropriately chosen wave function is always greater than the energy for the exact wave function. A frequently chosen definition of the quality of the wave function is how close the predicted variational energy comes to the exact energy. Though the exact energy is generally not known, the lower energy is guaranteed by the variational theorem to be closer to the exact answer. The lower the energy predicted for a wave function, the better the wave function is considered to be. Response **(C)** is correct, since minimizing the variational energy with respect to a variable parameter will lead to be a better wave function.

EST-2. In molecular mechanics studies of molecular structure, which has the largest force constant resisting structural change in a molecule?

(A) torsion angle twist (B) bond angle bend
(C) bond length stretch (D) change in position of the center of mass

Knowledge Required: The basic terms included in a molecular mechanics force field; the relative energies of different modes of molecular vibrations.

Thinking it Through: A change in position in the center of mass corresponds to translation of the molecule. Molecular mechanics does not consider translational motion. Response **(D)** is therefore incorrect. The magnitudes of the force constants for molecular vibrations occur in the order torsion angle twist < bond angle bend < bond stretch. Bond stretches have the largest force constants; response **(C)** is correct. Note that MM force fields generally don't treat torsions in terms of Hooke's law and a force constant!

EST-3. Hartree-Fock molecular orbital calculations are said to be an "SCF" type method. SCF stands for

(A) solutions by complex factors (B) semi-empirically calculated functions
(C) should contain *f*-orbitals (D) self-consistent field

Knowledge Required: Acronyms used to describe quantum mechanical calculational methods.

Thinking it Through: SCF stands for self-consistent field. Response **(D)** is correct.

EST-4. The exact ground state energy of He is –79.0 eV. Using the variation method, you calculate an approximate energy to be –83.0 eV. You must have made an error because variation method energies must

(A) lie above the ground state
(B) be positive
(C) equal the exact ground state energy
(D) be at least twice the exact ground state energy

Knowledge Required: Application of variation method to real calculations. Relative energies for calculations and true energies.

Thinking it Through: The variation method can be expressed through this variation theorem which states

$$E_{act(true)} \leq E_{trial} = \frac{\int \psi_{trial}^* \hat{H} \psi_{trial} d\tau}{\int \psi_{trial}^* \psi_{trial} d\tau}$$

Comparing the two energies given in the problem, −79.0 and −83.0, we see that the trial energy is less than (more negative) than the true energy violating the variation theorem. The trial energies must be less negative than the true energies; response **(A)** must be the correct response.

EST-5. The first three vibrational energy levels are marked A, B, and C on the diagram for the molecular potential energy curve for a diatomic molecule. From this information, the best estimate for the vibration frequency of the fundamental transition is

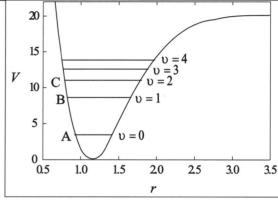

(A) $(B − A) / h$ **(B)** $(C − B) / h$ **(C)** $(C − A) / h$ **(D)** $(B − A) / 2h$

Knowledge Required: The selection rules for vibrational spectral transitions; the Bohr frequency condition, $\Delta E = h\nu_{obs}$, where ΔE is the energy difference between two quantum states. The $\upsilon = 0 \rightarrow \upsilon = 1$ transition is called the fundamental while other transitions from the $\upsilon = 0$ state are called overtones.

Thinking it Through: We can eliminate response **(C)** because it corresponds to an overtone. Response **(B)** can also be eliminated because although it satisfies the selection rule it is not the fundamental transition. Response **(D)** is incorrect because of the factor of 2 in the denominator which is not present in the Bohr frequency condition. Thus the correct response is **(A)**.

EST-6. Configuration interaction calculations account for

(A) correlation of electron motion

(B) correlation of nuclear motion

(C) shielding of a nucleus by inner shell electrons in a molecule

(D) the momentum increase of electrons in a molecular orbital as they approach the nucleus

Knowledge Required: The terms involved in calculated methods.

Thinking it Through: Configuration interaction calculations include multiple determinants in an attempt to better describe the motion of electrons. The motion of nuclei is associated with the Born-Oppenheimer approximation; response **(B)** is incorrect. Response **(C)** is not the best answer because configuration interaction calculations account for more than inner shell electrons. Response **(D)** is incorrect because the momentum changes throughout the description. Thus response **(A)** must be correct.

Electronic Structure Theory and Spectroscopy

EST-7. Molecular mechanics calculations use
- (A) a classical model based on the harmonic oscillator.
- (B) an LCAO-MO approach using hydrogenic orbitals.
- (C) a molecular orbital approach which is calibrated by experimental values.
- (D) a molecular orbital approach which uses a very large basis set but is not calibrated with experimental values.

Knowledge Required: Differentiation between calculational methods for molecules.

Thinking it Through: Molecular mechanics relies on a classical mechanics description of the forces within a molecule. These forces are described by the harmonic oscillator model; thus response **(A)** is correct. The other three responses correctly describe other approaches to calculations on molecules.

EST-8. The intensity of one of the vibrational absorption bands for a gaseous diatomic molecule is observed to increase with increasing temperature. This transition is likely to be

| (A) $\upsilon = 0 \to \upsilon = 1$ | (B) $\upsilon = 0 \to \upsilon = 2$ | (C) $\upsilon = 1 \to \upsilon = 2$ | (D) $\upsilon = 0 \to \upsilon = 3$ |

Knowledge Required: Temperature dependence of spectroscopy.

Thinking it Through: Three of the four responses involve transitions involving $\upsilon = 0$. The other involves $\upsilon = 1$, which is the only state which will show a significant increase in population with increasing temperature; thus response **(C)** is correct. Note that responses **(B)** and **(D)** correspond to overtones, which should be forbidden under harmonic oscillator selection rules.

EST-9. For an allowed electric dipole transition between the vibrational states 1 and 2 for a diatomic molecule with the z axis along the internuclear axis, which integral expression must be non-zero?

- (A) $\int \psi_2^* z \psi_1 \, d\tau$
- (B) $\int \psi_2^* \psi_1 \, d\tau$
- (C) $\int \psi_2^* z^2 \psi_1 \, d\tau$
- (D) $\int \psi_2^* z \psi_2 \, d\tau$

Knowledge Required: The transition moment integral for allowed spectra.

Thinking it Through: The transition moment integral is of the form given by response **(A)** which is correct. Response **(B)** fails to include the $\hat{z}$ operator. Response **(C)** contains the $\hat{z}^2$ operator, which is incorrect. Response **(D)** has the same function ψ_2 in the integral.

EST-10. The molecular orbital wave function for H₂ is proportional to the function $\varphi_a(1)\varphi_a(2) + \varphi_b(1)\varphi_b(2) + \varphi_a(1)\varphi_b(2) + \varphi_a(2)\varphi_b(1)$

where φ_a and φ_b are atomic orbitals for electrons localized on hydrogen atoms A and B, respectively. The set of valence bond structures appearing in this wave function is

- (A) $H_A^+H_B^-$, $H_A^-H_B^+$, and H_AH_B.
- (B) $H_A^+H_B^-$ and $H_A^-H_B^+$.
- (C) H_AH_B.
- (D) the set in **(A)** plus additional valence bond structures.

> **Knowledge Required:** Nature of terms in a molecular orbital wave function.
>
> **Thinking it Through:** The first term corresponds to both electrons on atom a, the second term to both electrons on atom b, and then remaining terms correspond to one electron on each atom. Thus response **(A)** is the correct.

Practice Questions

1. When the variation theorem is employed to estimate the energy for a system, the energy, E, is calculated using a trial wave function, φ. If E_0 is the correct energy of the ground state, then the energy calculated using ψ is

   (A) $E \geq 0$ (B) $E \geq E_0$
   (C) $E = E_0$ (D) $E \leq E_0$

2. Which basis set will result in the Hartree-Fock calculation having the lowest energy?

   (A) STO–3G (B) 3–21G
   (C) 6–31G   (D) 6–31G*

3. At room temperature the vibrational transition in HBr (gaseous) from the $\upsilon = 1$ state to the $\upsilon = 2$ state is much less intense than that for the transition from the $\upsilon = 0$ state to the $\upsilon = 1$ state. The primary reason for this is that

   (A) the former transition is forbidden while the latter is allowed.
   (B) the dipole moment of the $\upsilon = 0$ state is larger than that for the $\upsilon = 1$ state.
   (C) the Boltzmann thermal distribution is unfavorable.
   (D) the $\upsilon = 0$ state has more rotational states than the $\upsilon = 1$ state.

4. For two atomic orbitals to combine favorably to form a bonding molecular orbital in a heteronuclear diatomic molecule where the z axis is the internuclear axis, they must have identical

   (A) principal atom quantum numbers and about the same energy.
   (B) z components of orbital angular momentum and about the same energy.
   (C) orbital angular momenta and about the same energy.
   (D) energies.

5. The spatial portion of the molecular orbitals for H_2 can be expressed as
   $\psi(1,2) = c_1 \{1s_A 1s_B - 1s_B 1s_A\} +$
   $c_2\{1s_A 2s_B - 2s_A 1s_B\} +$
   $c_3 \{1s_A 2s_B + 2s_A 1s_B\} +$
   $c_4 1s_A 1s_A + \ldots$
   In the Hartree – Fock expression for the wavefunction for the bound molecule,

   (A) the term containing c_4 dominates for the equilibrium internuclear distance.
   (B) $c_2 = c_3 = c_4 = 0$.
   (C) $c_2 = c_3$.
   (D) the term containing c_1 corresponds to an ionic state.

6. The spatial portion of the molecular orbitals for H_2 can be expressed as
   $\psi(1,2) = c_1\{1s_A 1s_B - 1s_B 1s_A\} +$
   $c_2\{1s_A 2s_B - 2s_A 1s_B\} +$
   $c_3 \{1s_A 2s_B + 2s_A 1s_B\} +$
   $c_4\ 1s_A 1s_A + \ldots$
   In the Hartree – Fock expression for the wavefunction for the bound molecule,

   (A) the terms involving c_2 and c_3 are excited states.
   (B) the term containing c_1 represents an ionic state.
   (C) the c_2 term is a triplet state.
   (D) all are correct.

7. Which method of calculation is not variational?

   (A) Configuration Interaction
   (B) DFT
   (C) Hartree-Fock
   (D) MINDO

8. Which calculational method is most likely to result in the answer closest to the experimental heat of formation?

    (A) Hartree-Fock

    (B) density functional

    (C) molecular mechanics

    (D) depends on the molecule

9. Which calculation will likely take the longest for cyclopentadiene for the same starting conditions?

    (A) Hartree-Fock with 3-21G basis set

    (B) Hartree-Fock with 6-31G* basis set

    (C) B3LYP density functional calculation with 6-31G* basis set

    (D) MP2 calculation with 6-31* basis set

10. Which model system provides molecular orbitals composed of the fewest number of atomic orbitals for a diatomic molecule?

    (A) Gaussian orbitals

    (B) H_2^+ orbitals

    (C) Slater orbitals

    (D) symmetry adapted orbitals

Electronic Structure Theory and Spectroscopy

Answers to Study Questions

1. C
2. C
3. D
4. A
5. A
6. A
7. A
8. C
9. A
10. A

Answers to Practice Questions

1. B
2. D
3. C
4. B
5. B
6. D
7. B
8. D
9. D
10. B